Alexander König

Advancements and Future Directions in Rehabilitation Robotics

Alexander König

Advancements and Future Directions in Rehabilitation Robotics

Human-in-the-Loop Control during Neurological Gait Rehabilitation

Südwestdeutscher Verlag für Hochschulschriften

Impressum/Imprint (nur für Deutschland/only for Germany)
Bibliografische Information der Deutschen Nationalbibliothek: Die Deutsche Nationalbibliothek verzeichnet diese Publikation in der Deutschen Nationalbibliografie; detaillierte bibliografische Daten sind im Internet über http://dnb.d-nb.de abrufbar.
Alle in diesem Buch genannten Marken und Produktnamen unterliegen warenzeichen-, marken- oder patentrechtlichem Schutz bzw. sind Warenzeichen oder eingetragene Warenzeichen der jeweiligen Inhaber. Die Wiedergabe von Marken, Produktnamen, Gebrauchsnamen, Handelsnamen, Warenbezeichnungen u.s.w. in diesem Werk berechtigt auch ohne besondere Kennzeichnung nicht zu der Annahme, dass solche Namen im Sinne der Warenzeichen- und Markenschutzgesetzgebung als frei zu betrachten wären und daher von jedermann benutzt werden dürften.

Verlag: Südwestdeutscher Verlag für Hochschulschriften GmbH & Co. KG
Dudweiler Landstr. 99, 66123 Saarbrücken, Deutschland
Telefon +49 681 37 20 271-1, Telefax +49 681 37 20 271-0
Email: info@svh-verlag.de

Approved by: Zürich, Eidgenössische Technische Hochschule, Diss. 19641, 2011

Herstellung in Deutschland:
Schaltungsdienst Lange o.H.G., Berlin
Books on Demand GmbH, Norderstedt
Reha GmbH, Saarbrücken
Amazon Distribution GmbH, Leipzig
ISBN: 978-3-8381-2866-5

Imprint (only for USA, GB)
Bibliographic information published by the Deutsche Nationalbibliothek: The Deutsche Nationalbibliothek lists this publication in the Deutsche Nationalbibliografie; detailed bibliographic data are available in the Internet at http://dnb.d-nb.de.
Any brand names and product names mentioned in this book are subject to trademark, brand or patent protection and are trademarks or registered trademarks of their respective holders. The use of brand names, product names, common names, trade names, product descriptions etc. even without a particular marking in this works is in no way to be construed to mean that such names may be regarded as unrestricted in respect of trademark and brand protection legislation and could thus be used by anyone.

Publisher: Südwestdeutscher Verlag für Hochschulschriften GmbH & Co. KG
Dudweiler Landstr. 99, 66123 Saarbrücken, Germany
Phone +49 681 37 20 271-1, Fax +49 681 37 20 271-0
Email: info@svh-verlag.de

Printed in the U.S.A.
Printed in the U.K. by (see last page)
ISBN: 978-3-8381-2866-5

Copyright © 2011 by the author and Südwestdeutscher Verlag für Hochschulschriften GmbH & Co. KG and licensors
All rights reserved. Saarbrücken 2011

Acknowledgments

For Robert and Danie

Contents

Acknowledgments	i
List of Figures	vii
List of Tables	ix
List of Abbreviations	xi

1 Introduction   1
   1.1 Neurological Damage and Rehabilitation 1
   1.2 Active Participation and Motivation in Rehabilitation 2
        1.2.1 Active Physical Participation 2
        1.2.2 Cognitive Participation 2
        1.2.3 Motivation 2
   1.3 Virtual Environments for Modulation of Participation and Motivation in Rehabilitation 3
   1.4 Robots and Rehabilitation 4
   1.5 Integrating the Human into the Control Loop 4
        1.5.1 Human in the Loop Control 5
        1.5.2 Bio-Cooperative Control 6
        1.5.3 Biomechanical Human-in-the-Loop Control 6
        1.5.4 Physiological Human-in-the-Loop Control 7
        1.5.5 Psychological Human-in-the-Loop Control 8
                Modulating the Psychological State 8
                Detecting the Psychological State 8
                Previous Work 9
   1.6 Aims of This Thesis 10

2 Controlling Physiological States during Robot-Assisted Gait Training   11
   2.1 Introduction 11
   2.2 Methods 12
        2.2.1 Definition of Physical Effort 12
        2.2.2 Quantifying Physical Effort 13
                Physiological Quantification 13
                Biomechanical Quantification 13

Contents

	2.2.3 Controlling Physical Effort with Visual Instructions	14
	2.2.4 Controlling Heart Rate Using Treadmill Speed	15
	2.2.5 Experimental Protocols	16
	2.2.6 Controller Performance Evaluation	18
2.3	Results	20
	2.3.1 Control of HR via Treadmill Speed	20
	2.3.2 Control of HR via Visual Instructions	20
	2.3.3 Control of WIT via Visual Stimuli	22
	2.3.4 Statistical Comparison Between the Three Approaches	23
2.4	Discussion	25
	2.4.1 Patient Individual Control of Participation	25
	2.4.2 A Metric for Patient Individual Control of Physical Activity	26
	2.4.3 Cardiovascular Training After Stroke	27
	2.4.4 Clinical Applicability of Patient Activity Control	29
	Heart Rate	29
	WIT	29
2.5	Contribution	29

3 Controlling Psychological States during Robot-Assisted Gait Training   **31**

3.1	Introduction	31
3.2	Methods	33
	3.2.1 Common Methods	33
	Questionnaires	33
	Physiological Recordings	34
	Experimental Setup	36
	3.2.2 The Effects of Pleasant and Unpleasant Stimuli on Psychophysiological Recordings	36
	Virtual Task	36
	Experimental Protocol	36
	Subjects	37
	3.2.3 The Effect of Virtual Environments on Psychophysiological Recordings	38
	Virtual Task	38
	Inducing Different Levels of Cognitive Engagement	39
	Evaluation of Physiological Recordings	41
	Experimental Protocol	42
	Subjects	42
	3.2.4 Automatic Classification of Cognitive Load	45
	Input Data for Classification of Cognitive Load	45
	Virtual Task	45
	Modulating Cognitive Load	46
	Correlation between Physical Effort and Cognitive Load	47

		Setup of the Three Classifiers	47
		Experimental Protocol for Classifier Training	49
		Performance Evaluation	50
		Subjects	51
	3.2.5	Closed Loop Control of Cognitive Load	51
		Adaptation of Virtual Environment	52
		Experimental Protocol	53
		Performance Evaluation of Closed Loop Experiments	53
		Subjects	54
3.3	Results		55
	3.3.1	The Effects of Pleasant and Unpleasant Stimuli on Psychophysiological Recordings	55
		Questionnaires	55
		Physiological Recordings	55
	3.3.2	The Effect of Virtual Environments on Psychophysiological Recordings	57
		Questionnaires	57
		Statistical Analysis of Recorded Physiological Signals	57
		PCA Analysis of Recorded Physiological Signals	58
	3.3.3	Automatic Classification of Cognitive Load	63
		Physical Effort and Cognitive Load	63
		Classification Performance of Training Experiments	64
	3.3.4	Classifier Performance during Closed Loop Control of Cognitive Load	66
3.4	Discussion		68
	3.4.1	Pleasant and Unpleasant Stimuli in Psychophysiological Recordings	68
	3.4.2	The Effect of Virtual Environments on Psychophysiological Recordings	69
		Questionnaires	69
		Principle Component Analysis	69
		Study Limitations	70
	3.4.3	Classification Performance	71
	3.4.4	Necessity for Kalman Filters in Neurological Patients	71
	3.4.5	Transfer to Daily Clinical Routine	72
	3.4.6	The Influence of Physical Activity, Neurological Deficits and Medication	74
	3.4.7	Dissociation of Physical Effort and Cognitive Load	75
	3.4.8	Extension of the Approach to Upper Limbs	75
3.5	Contribution		76

Contents

4 Monitoring Heart Rate During Control of Physical Effort   **77**
4.1 Introduction 77
4.2 Methods 77
     4.2.1 Power Exchange during Lokomat Walking 77
     4.2.2 Model Identification in Healthy Subjects 79
     4.2.3 Model Identification in Patients 81
4.3 Results 83
     4.3.1 Model Identification 83
     4.3.2 Heart Rate Prediction of Healthy Subjects and Patients . . . 84
4.4 Discussion 88
4.5 Contribution 88

5 Conclusions and Outlook   **89**
5.1 Key Findings 89
5.2 The Impact of Human in the Loop Control on Rehabilitation Robotics 90

Bibliography   **91**

List of Figures

1.1 The Lokomat Gait Orthosis 5
1.2 The Human-in-the-loop control scheme 7

2.1 Virtual scenario used for control of WIT and heart rate 15
2.2 Control scheme for control of activity via a visual stimulus 16
2.3 Control scheme for PI control of HR via treadmill speed 17
2.4 Results of PI control of HR via adaptation of treadmill speed 20
2.5 Results of HR control via visual instructions 21
2.6 Results of WIT control via visual instructions 23
2.7 Boxplots comparing the three different approaches to effort control . 24
2.8 Selection matrix for optimal training of stroke patients, depending on the patient's cognitive and physical impairments 27

3.1 Three conditions were defined according to the circumplex model of affect 34
3.2 Pictorial questionnaire on physical effort 34
3.3 ECG beat detection algorithm 35
3.4 The multimodal controller setup 37
3.5 Virtual environment with pleasant, unpleasant and scary stimuli . . 38
3.6 The virtual task 39
3.7 Inducing different states of cognitive engagement during gait training using the task level difficulty of a virtual reality task 40
3.8 Experimental protocol 43
3.9 Virtual scenario 46
3.10 Study protocol open loop experiments 50
3.11 Adaptation of virtual environment based on the result of the classifier 52
3.12 Variance explained by PCs for healthy subjects and patients 58
3.13 First two activation coefficients of the PCA exemplarily shown for one healthy subject (subject ID 17) 59
3.14 First two activation coefficients of the PCA exemplarily shown for one patient (patient ID 3) 60
3.15 Comparison between the loading factors of the first three PCs of healthy (n=17) subjects and patients (n=10) 64
3.16 Boxplots of perceived physical effort and cognitive load in healthy subjects and patients 67

List of Figures

3.17 Exemplary plot from data of healthy subject 4 from closed loop control of cognitive load using the KALDA classifier with physiological input 73

4.1 Forces/torques and corresponding power exchange between Lokomat and patient 78
4.2 Simulink model of the heart rate model components 83
4.3 Changes in heart rate for different settings of treadmill speed and different levels power expenditure 84
4.4 Predicted and recorded heart rate of one healthy subject subject and Lokomat treadmill velocity profile for model verification 87

List of Tables

2.1	Overview over the experiments performed	17
2.2	Characteristics of patients of HR and WIT control experiments	19
2.3	Minimum and maximum HR values of each patient used for control of HR via treadmill speed	21
2.4	Minimum and maximum HR values of each patient used for control of HR via visual instructions	22
2.5	Minimum and maximum WIT values of each patient used for control of WIT via visual instructions	22
3.1	Characteristics of patients	44
3.2	Characteristics of patients for open loop classifier training	51
3.3	Characteristics of patients for closed loop experiments	54
3.4	Changes in physiological signals caused by pleasant and unpleasant stimuli	56
3.5	Results of healthy subjects for the self-assessment manikin questionnaire (SAM)	57
3.6	Statistical results of physiological recordings in healthy subjects	61
3.7	Loading factors of the first PCs of subject 17	62
3.8	Loading factors of the first PCs of patient 3	63
3.9	Open loop, "leave one out" classification results for healthy subjects and patients for four different kinds of input vectors	65
3.10	Results of closed loop experiments in healthy subjects and patients	66
4.1	Patient data from model identification recordings. (©2009 IEEE)	82
4.2	Results of changes in guidance force for all patients in experiment I	85
4.3	Results of changes in for all patients in experiment II	86
4.4	Prediction quality of the heart rate model with optimization of 4 parameters	86

List of Abbreviations

ANS	Autonomic Nervous System
CI	Confidence Interval
CNS	Central Nervous System
DGO	Driven Gait Orthosis
ECG	Electrocardiogram
EEG	Electroencephalogram
EMG	Electromyogram
HR	Heart Rate
HRV	Heart Rate Variability
KALDA	Kalman Adaptive Linear Discriminant Analysis
LDA	Linear Discriminant Analysis
PC	Principal Component
PCA	Principal Component Analysis
PI	Proportional Derivative
RMSE	Root Mean Square Error
RMSSD	Square Root of the Mean Squared Differences of Successive Normal-to-Normal Intervals
SAM	Self-Assessment Manikin
SCI	Spinal Cord Injury
SCL	Skin Conductance Level
SCR	Skin Conductance Responses
SD	Standard Deviation
v_{TM}	Treadmill Speed of the Lokomat
VR	Virtual Reality
WIT	Weighted Interaction Torques

1 Introduction

1.1 Neurological Damage and Rehabilitation

The Central Nervous System (CNS) is capable of generating an amazing variety of movements with the lower extremities, ranging from walking and running over jumping to highly skilled object manipulations such as soccer ball tricks. We can acquire new motor skills or hone existing ones by exercising and practicing these skills. Motor learning thereby emerges from the ability of the CNS to reorganize itself. The underlying mechanism is called neural plasticity (Nudo, 2006) and refers to the ability of the CNS to make new connections between neurons and strengthen existing ones, thereby constantly changing its structure.

Damages of the CNS, such as stroke or Spinal Cord Injury (SCI), are amongst the leading causes of disabilities, severely limit the quality of life of affected people and their possibility to actively contribute to society (Anderson, 2004). Recent studies estimate the incidence of stroke to at least 101 - 285 in men and 47 - 198 in women per 100.000 subjects in Europe (Thorvaldsen et al., 1995; Brainin et al., 2000). Thus, stroke affects about 1 million people in Europe each year. SCI is reported to affect 14-20 subjects per million in Europe (Wyndaele and Wyndaele, 2006) and 14-40 subjects per million worldwide (Sekhon and Fehlings, 2001).

Motor rehabilitation exploits the concept of plasticity and can improve the impairment and thereby the standard of living of neurological patients. SCI patients reported the inability to walk to be one of the most important impairments that limited their quality of life, besides bladder/bowel control, sexual dysfunction and loss of proprioception (Anderson, 2004). In order to improve existing but limited walking capabilities, treadmill training has become the gold standard in rehabilitation programs administered to stroke or incomplete SCI patients (Shumway-Cook and Woollacott, 1995; Kwakkel et al., 1997; Dietz et al., 1998; Dietz and Duysens, 2000; Kwakkel et al., 2002; Cherng et al., 2007).

1 Introduction

1.2 Active Participation and Motivation in Rehabilitation

1.2.1 Active Physical Participation

Active physical participation of the patients in rehabilitation training and high training intensities were shown to be an important factor for successful rehabilitation (Langhorne et al., 1996; Kwakkel et al., 1997, 2002; Lotze et al., 2003). High voluntary physical effort and active contribution in a movement were shown to improve motor learning and rehabilitation (Lotze et al., 2003; Kaelin-Lang et al., 2005). Heart rate (HR) is a widely used measure for physical activity. During gait rehabilitation of stroke survivors, cardiovascular training can be of great benefit to the patient's rehabilitation outcome (Gordon et al., 2004), as it prevents secondary complications. Up to now, non-ambulatory patients cannot exercise on treadmills, but must use ergometers such as stationary bicycles, where the problems of coordination and balance during walking cannot be taken into consideration.

1.2.2 Cognitive Participation

In addition, active cognitive participation and motivating training sessions are key requirements for the success of motor learning in general and in rehabilitation (Maclean and Pound, 2000a; Loureiro et al., 2001; Lotze et al., 2003; Kaelin-Lang et al., 2005; Liebermann et al., 2006). The learning rate of a motor task is maximal at a task difficulty level that positively challenges and excites subjects while not being too stressful or boring (Guadagnoli and Lee, 2004). Research in healthy subjects suggests that motor learning decreases in the presence of a distracting cognitive task, which presents a cognitively over-challenging situation (Redding et al., 1992; Taylor and Thoroughman, 2008). A task which is too easy for the subject will be perceived as boring, a task which is too difficult will overstress the subject, while an optimally challenging task should induce maximal motivation and cognitive participation.

1.2.3 Motivation

Motivation of patients to actively participate in the therapy, with physical effort and with cognitive participation, can be regarded as the most important factor for successful rehabilitation (O'Gorman, 1975). In the context of neurological rehabilitation, motivation can be defined as the direction and intensity of one's effort (Sage, 1977), or as the forces acting on or within a person to initiate a behavior (Phillips et al., 2004). Motivation is seen as a prerequisite of rehabilitation success (Maclean and Pound, 2000b), particularly as active physical and cognitive participation towards the training intervention is usually equated with high motivation, and passivity with lack of motivation (Colombo et al., 2007). Assessment of motivation is difficult, and typically done via questionnaires such as the Intrinsic Motivation Inventory (IMI) (McAuley et al., 1989).

1.3 Virtual Environments for Modulation of Participation and Motivation in Rehabilitation

Motivation is, however, not only difficult to assess, but the possibilities of modulating motivation are limited. Virtual environments are commonly used to modulate motivation, challenge patients to longer training duration and cadence (Liebermann et al., 2006) and to modulate patient participation (Holden, 2005; Mirelman et al., 2009; Brütsch et al., 2010). The patient can obtain intuitive and easy to understand information on his or her performance during the training (Banz et al., 2008). Virtual environments are used as an information tool to update the patient about his or her range of motion and maximal force (Girone et al., 2000) or quality of gait movements (Banz et al., 2008). These virtual environments can be therapeutically superior to real scenarios (Sveistrup, 2004; Holden, 2005). Virtually enriched environments as well as functional and task-oriented exercise environments were shown to improve motor re-learning and recovery after stroke (Johnson, 2006). Virtual reality can be used to test different types of feedback such as congruent or terminal feedback for comparative effectiveness in improving motor function in patients. Virtual reality technology thereby provides a convenient mechanism for manipulating these factors, setting up automatic training schedules and for training, testing, and recording participants' motor responses.

A study in healthy subjects showed that virtual reality enabled healthy subjects to perform more accurate movements during obstacle stepping (Wellner et al., 2008). Subjects were given auditory, visual and haptic feedback on their foot clearance and the distance to obstacles they had to overstep. Performance was measured as foot clearance over the obstacle and the number of obstacles hit. This study showed that subjects had higher performance with auditory and visual feedback than with visual feedback alone. Additionally, the authors showed that 3D vision did not improve performance compared to 2D vision.

In patients, a study by Brütsch showed that virtual reality had the potential to increase active participation of children with cerebral palsy (Brütsch et al., 2010) compared to gait therapy alone. Participation was thereby quantified by EMG measurements. Mirelman showed in a randomized controlled trial study that the use of VR increased the usage of a home-based ankle rehabilitation system after stroke (Mirelman et al., 2009). The increased time for the exercise significantly improved the functional recovery compared to patients that exercised without VR,

1 Introduction

measured as gait speed and distance walked. Working with SCI patients, Ohsuga concluded that immersion in VR might be expected to temporarily reduce pain and relieve anxieties (Ohsuga et al., 1998). Fung used a VR system coupled to an omnidirectional treadmill, in which stroke patients could exercise adaptation of their gait speed in a realistic environment (Fung et al., 2006).

Recently, virtual environments and video games have also been connected to increased neural plasticity in the CNS, as changes on a molecular and cellular level seem to be induced by video games (Bavelier et al., 2010). These results are however preliminary, and a clear influence of virtual reality on neural plasticity could not yet be established beyond the effects induced by increased physical and cognitive participation via multi-modal feedback.

1.4 Robots and Rehabilitation

Robots have become increasingly common to automate rehabilitative treadmill training, as they allow for longer training duration and higher training intensity (Wirz et al., 2005). Two general design approaches have been pursued: end-effector based robots such has the Gaittrainer (Hesse et al., 1999), or the Haptic Walker (Schmidt et al., 2007) and exoskeleton robots such as the Lokomat displayed in Figure 1.1 (Colombo et al., 2000; Riener et al., 2010), the Lopes (Veneman et al., 2007), the Autoambulator (www.healthsouth.com) or the Walk Trainer (Stauffer et al., 2009).

Despite their advantages, the effectiveness of gait rehabilitation robots is still discussed controversially. Recent reviews of Mehrholz (Mehrholz et al., 2007, 2008) could not identify a clear advantage of gait rehabilitation robots over conventional physiotherapy. The Lokomat, as the most intensively studied amongst all available gait robots, was found to be superior to manual therapy (Mayr et al., 2007; Schwartz et al., 2009), equally efficient as manual therapy (Westlake and Patten, 2009) or inferior to manual therapy (Hidler and Wall, 2005; Hornby et al., 2008).

1.5 Integrating the Human into the Control Loop

A possible explanation, why some studies found gait robots to be inferior compared to manual therapy (Hidler and Wall, 2005; Hornby et al., 2008), might be that the whole training environment does not yet react in a bio-cooperative way. Bio-cooperative behavior is defined as the ability of the robot to react compliantly to the user's voluntary effort and intention (Riener et al., 2005b, 2009a,b; Riener and Munih, 2010), thereby integrating the human in the control loop instead of treating the human as a source of perturbation.

1.5 Integrating the Human into the Control Loop

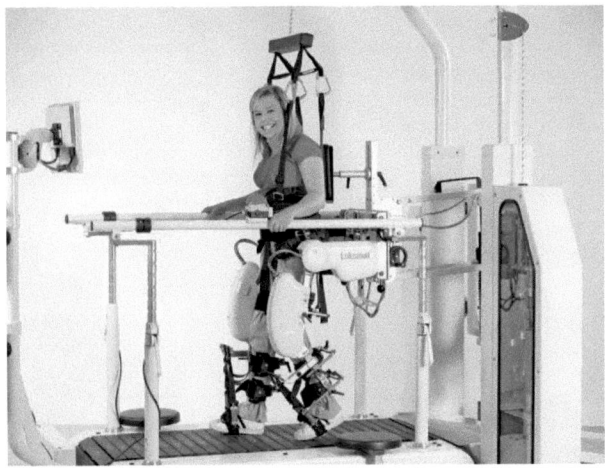

Figure 1.1: The Lokomat Gait Orthosis (Image courtesy: Hocoma AG, Volketsvil, Switzerland)

1.5.1 Human in the Loop Control

Historically, human-in-the-loop control refers to the concept of a (mechanical) coupling between human and robot for direct interaction of the human with the robot's end-effector. The development of robots has been driven by the need of performing repetitive movements at high precision and accuracy in a well structured and known environment, as for example in the car manufacturing industry. The robot was position-controlled and humans had to avoid interaction with the robot to ensure safety and to not introduce disturbances.

First controllers that accounted for the human in the loop were introduced as impedance control by Hogan (Hogan, 1985), who provided the theoretical basis of compliant robotics which allow users to interact with the robot and deviate from the predefined position trajectory. Interaction with the environment being the fundamental basis of manipulation (Colgate and Hogan, 1989), robotic arms need to react in a safe, mechanically compliant, flexible, gentle and adaptive way to the user (Riener et al., 2005a). These user-cooperative interactions are highly challenging, when human and robot are in direct contact with each other or share the same workspace, as the environment the robot has to interact with might not be completely known, might be variable in it's shape and appearance and become unstable due to energy introduced into the system by the human (Adams et al., 1998).

1 Introduction

A large body of research in human in the loop control is focused on controllers that allow stable haptic interaction of humans in a virtual environments (Buttolo et al., 1997). The robot is used to provide haptic feedback to the user and make the virtual experience more realistic by augmenting acoustic and visual feedback with haptic feedback (Adams and Hannaford, 1999). Main fields of application are teleoperation (Fong et al., 2003), surgical robotics (van der Meijden and Schijven, 2009) and training and simulation environments. These simulation environments such as medical training simulations (Basdogan et al., 2001) allow exercising complex and dangerous tasks in a safe environment, in which clinicians can exercise procedures on a virtual patient model.

1.5.2 Bio-Cooperative Control

In the following, bio-cooperative behavior refers to the concept of not only integrating the user on a mechanical, but also on a biomechanical, physiological and psychological level (Riener et al., 2005b, 2009a,b; Riener and Munih, 2010). In the context of bio-cooperative control, the rehabilitation robot therefore needs to react compliantly to the patient's needs and demands.

On the biomechanical level, the biomechanics of the human are being treated cooperatively, and the robot reacts compliantly and adaptively to the biomechanical capabilities of the patients (Figure 1.2).

On the physiological level, the robot has to take the physiology of the patient into account and needs to react adaptively to changing demands in physical effort.

On the psychological layer, the robot has to interact with the patient on a cognitive level and take the patient's current motivation, mental engagement or cognitive load into account (Riener et al., 2009a,b; Riener and Munih, 2010). Following Riener et al. (Riener et al., 2009a,b; Riener and Munih, 2010), I define cognitive load as the amount of focus and concentration the patient has to invest to still be able to fulfill the rehabilitation task.

1.5.3 Biomechanical Human-in-the-Loop Control

On the biomechanical level, recent advances in cooperative control strategies allow rehabilitation robots to integrate the patient into the control loop on a biomechanical level (Duschau-Wicke et al., 2010). While first rehabilitation robots were position controlled and imposed a predefine gait trajectory on the patient (Colombo et al., 2000), impedance control allowed the patient to deviate spatially from the robot's trajectory (Hogan, 1985; Riener et al., 2005b). The "path control" approach (Duschau-Wicke et al., 2010), implemented in the Lokomat gait orthosis, can react adaptively to the patient's abilities by learning a force field that supports the gait movement depending on the patient's impairments.

1.5 Integrating the Human into the Control Loop

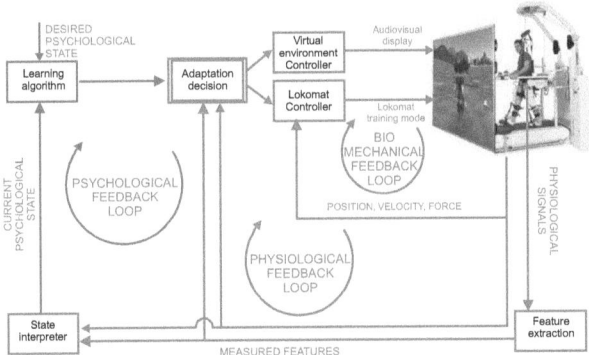

Figure 1.2: The Human-in-the-loop control scheme. The human is part of a closed loop control of biomechanics (blue), physiology (orange) and psychology (green)

1.5.4 Physiological Human-in-the-Loop Control

Current rehabilitation environments still lack cooperative behavior towards the physiological integration of the patient into the control loop (Riener et al., 2009a,b), although active contribution in a movement was shown to be crucial for motor learning and rehabilitation (Lotze et al., 2003; Kaelin-Lang et al., 2005). Gait robots are strong enough to move the patient's legs along a predefined walking trajectory and can support learned passiveness of the patient. Motivating the patient to active participation in the robot can be seen as a key factor to improve the success of gait robots (Hidler and Wall, 2005; Israel et al., 2006). A lack of active participation might explain the inconclusive effect of rehabilitation robots as shown in studies from Israel et al. and Hidler and Wall, who found decreased muscle activity for robot-assisted walking compared to non assisted walking.

In the light of the literature review in section 1.2, the combination of gait training with physical endurance training can be regarded as key to improvement of neurological rehabilitation. Coordinative gait training on the one side play a major role in rehabilitation of stroke survivors (Shumway-Cook and Woollacott, 1995). Depending on the degree of impairments caused by the lesion, this training is performed either on treadmills for less severe cases or on stationary bicycles in severely affected patients. Particularly non-ambulatory patients cannot exercise on treadmills, but must use stationary bicycles, where the problems of coordination and balance during walking do not need to be taken into consideration. Intense exercise of physical endurance on the other side was shown to improve sensorimotor functions, decrease cardiovascular risk factors and improve the medical risk man-

1 Introduction

agement of stroke survivors (Gordon et al., 2004). The Clinical Council of the American Heart Association sees physical endurance exercise to be an integral part of rehabilitation after neurological injury and suggests that it should play a central role in stroke rehabilitation (Gordon et al., 2004).

At the same time, monitoring heart rate during exercise can be crucial to prevent over-training, which would decrease the efficacy of the rehabilitation training (Achten and Jeukendrup, 2003) and might become dangerous to the patient.

1.5.5 Psychological Human-in-the-Loop Control

Modulating the Psychological State

Although tools exist to modulate the psychological state of a patient such as virtual environments (Holden, 2005; Banz et al., 2008; Mirelman et al., 2009; Brütsch et al., 2010), current rehabilitation environments are not programmed to adapt to the psychological patient states. One major reason is that the current psychological cognitive load of patients cannot be objectively assessed. Questionnaires such as the "Intrinsic Motivation Inventory" (McAuley et al., 1989) can be used to obtain subjective information, but only at discrete time-points after training has ceased. They can therefore not be used in real time. Questionnaires can therefore not be used in real time. In addition, neurological patients with severe cognitive deficits or aphasia might not be able to understand and respond appropriately to the questions.

Detecting the Psychological State

Psychophysiological measurements can provide real-time information on the cognitive load of subjects (Andreassi, 2007; Duffy, 2008), as physiological processes were shown to reflect behavioral-, cognitive-, emotional- and social interaction (Hugdahl, 1995). The physiological recordings reflect reactions that result from activities of the CNS and the Autonomic Nervous System (ANS). Examples for measures of the CNS are the electroencephalogram (EEG) or near infrared spectroscopy (NIRS), examples for measurements of the ANS are electrocardiogram (ECG), galvanic skin response, breathing frequency or skin temperature. Measuring brain activity via EEG showed that increased cognitive load caused a decreased alpha wave activity (8-12 Hz) and an increase in beta wave activity (≥ 13 Hz) (Duffy, 2008). However, real-time measurement and analysis of signals from the CNS during walking in a robotic device are in general not feasible due to noise and motion artifacts. Up to now, psychological state detection therefore focused on signals from the ANS.

Signals from the ANS that reflect the psychological state of a human are primarily signals that respond to mental stress or relaxation (Andreassi, 2007). In addition to psychological processes, physical effort, such as walking on a treadmill, can influence the psycho-physiological measurements. From the ECG, HR and heart rate variability (HRV) can be computed. When recorded during a virtual task with a goal, HR was shown to be an indicator of physical as well as cognitive load (Mulder et al., 2000). Physiological effort and psychological stress have an influence on the short-term variation of HR. HRV was shown to decrease during physical effort (de la Cruz Torres et al., 2008), mental stress (Delaney and Brodie, 2000) and negative emotions (McCraty et al., 1995). Galvanic skin response is used as a direct measure for arousal (Ax, 1953; Cacioppo et al., 2000) and was found to increase during demanding tasks compared to a rest period (Dawson et al., 2007). From the galvanic skin response, skin conductance responses (SCR) measured as a number, and the skin conductance level (SCL) are computed. The number of SCR is a sensitive indicator for emotional strain (Boucsein, 2005). In recent research, SCL was found to increase during demanding tasks compared to a rest period (Dawson et al., 2007). The breathing frequency was found to increase during stress (Suess et al., 1980), negative emotions (Boiten et al., 1994) and cognitive effort (Carroll et al., 986b) and also during physical activity (Rousselle et al., 1995).

However, physiological signals that provide information on cognitive load can be ambiguous or even contradictory. HR was found to increase due to stress or negative emotions (Levenson et al., 1990; Andreassi, 2007), but decreased in reaction to unpleasant stimuli (Palomba et al., 2000; Gomez et al., 2005; Codispoti et al., 2008). Skin temperature decreased during cognitive load in a study by Ohsuga (Ohsuga et al., 2001) but increased with physical activity (Mancuso and Knight, 1992). Other physiological recordings from the peripheral nervous system have been used as indicators of the psycho-physiological state of a subject. Amongst these were facial EMG recordings as indicators for emotional responses to pleasant or unpleasant stimuli (Brown and Schwartz, 1980; Larsen et al., 2003).

Previous Work

Previously, mental engagement of neurological patients has been automatically quantified during robot assisted rehabilitation using psychophysiological signals (Novak et al., 2010). Closed loop control of psychological states in healthy subjects has been implemented to adjust the difficulty level or level of assistance in virtual tasks. Haarmann et al. (Haarmann et al., 2009) performed a study in 48 healthy subjects and combined GSR with HRV measurement to control the difficulty level of a flight simulation task. Also in an aviation task, Wilson et al. (Wilson and Russell, 2007) adapted the level adaptive assistance depending on a psychophysiological estimation of a subject's workload. Rani et al. (Rani et al., 2002) estimated stress from HRV using a Fuzzy classifier and controlled stress to a desired level. Liu et al. (Liu et al., 2009) used physiological signals to adapt computer game difficulty in real-time.

1 Introduction

1.6 Aims of This Thesis

Based on the literature summarized in the previous sections, it can be concluded that controlling physical effort and cognitive load of a patient to avoid tasks that are physically and cognitively too demanding or too easy has the potential to increase motor learning and thereby the training efficiency and therapeutic outcome of neurological rehabilitation. Psychophysiological measurements in combination with virtual environments can be used to perform bio-cooperative control by putting the human in a psychological closed control loop (Riener and Munih, 2010). During these experiments, heart rate should be monitored in order to prevent overexertion.

The aim of this thesis was to perform bio-cooperative control on physical effort and cognitive load of patients during robot-assisted gait training in a stable, reliable, and easy-to-use manner while guaranteeing patient safety at all times. The proposed solutions had to cover patients with a broad variety of physical as well as cognitive impairments and adapt automatically to the patients' specific demands and needs. The Lokomat gait orthosis was used for all experiments, but the approaches presented in this work are generalizable to any gait robot equipped with force sensors.

In chapter 2, several different solutions for control of physical effort to a desired temporal profile are described. Possible physiological and non-physiological markers are identified that reflect physical effort. Depending on the patient's physical and cognitive impairments, control strategies are derived and exemplarily evaluated in healthy subjects, stroke and SCI patients. The chapter concludes with a metric that will allow selecting the solution best suitable for the patient's demands in terms of physical and cognitive impairment.

In chapter 3, control of cognitive load in patients and healthy subjects is described. First, cognitive load is defined in the context of gait rehabilitation in stroke subjects. Using virtual environments, cognitive load is then modulated during robot-assisted gait training. Psychophysiological signals are used to objectively estimate the current cognitive load in real-time via linear and nonlinear classification techniques. Performance measures as a proxy from physiological signals are then investigated. Finally, the clinical applicability is discussed and the possible impact is evaluated, which controlling cognitive load could have on the field of neurological rehabilitation.

In chapter 4, a model capable of foreseeing the temporal evolution of the heart rate is developed. This model can predict when the heart rate is likely increase over a security threshold caused by the physical stress of the exercise. The prediction quality of this model is then tested in healthy subjects and patients.

The thesis ends with a summary of the key findings and general conclusions on the impact of bio-cooperative control on neuro-rehabilitation in chapter 5 .

2 Controlling Physiological States during Robot-Assisted Gait Training

2.1 Introduction

Active contribution in a movement was shown to be crucial for motor learning and rehabilitation (Lotze et al., 2003; Kaelin-Lang et al., 2005). As gait robots are strong enough to move the patient's legs along a predefined walking trajectory, active participation of the patient can be seen as a key factor to improve the success of gait robots.

A lack of active participation might explain the inconclusive effect of rehabilitation robots, as subjects can behave passively in the robot as shown in studies from Israel et al. (Israel et al., 2006) and Hidler et al. (Hidler and Wall, 2005), who found decreased muscle activity for robot-assisted walking compared to non assisted walking. On a biomechanical level, cooperative "assist-as needed" controllers can promote active participation (Duschau-Wicke et al., 2010). On a cognitive level, visual feedback was shown to help patients to focus on their walking movement (Banz et al., 2008). Virtual environments were shown to improve motivation of patients (Holden, 2005; Brütsch et al., 2010) and increased rehabilitation success (Mirelman et al., 2009).

However, there is no effective method for controlling patient participation during robot assisted gait training to a desired level. Due to the broad variety of physical and cognitive impairments of stroke patients, a "one-size fits all" solution for control of patient participation is unlikely to suit the demands of all patients. In particular, severe cognitive impairments limit the ability of the patient to understand which movements are recommended by the therapist and which movements are beneficial for therapeutic success.

In this chapter, several approaches to controlling patient participation during robot-assisted gait therapy are presented. First, patient activity was quantified in the presence of a gait robot. Heart rate (HR) and interaction forces between robot and patient were used as indicators of patient activity. Then, two methods are derived to control the measured of patient activity: control via treadmill speed adaptations and control via visual feedback. In three experiments, the controllers are tested in stroke patients. Finally, a metric is provided that allows selecting the solution that best suits the patient's demands in terms of physical and cognitive impairment. With this approach, an increase in activity during training compared to normal robot-assisted therapy could be expected which could have a beneficial effect on the rehabilitation outcome.

The methods and results presented in this chapter have been previously published in a journal paper (Koenig et al., 2011b) and a conference paper (Koenig et al., 2011d). The paper (Koenig et al., 2011b) first appeared at BioMed Central and is reprinted with permission.

2.2 Methods

To be able to control patient participation during robot-assisted walking, it was necessary to define and quantify the amount of participation. Patient participation was quantified in two ways: by HR, a physiological parameter that reflected physical effort during body weight supported treadmill training (Thomas et al., 2007) and by a weighted sum of the interaction torques (WIT) between robot and patient, recorded from hip and knee joints of both legs (Banz et al., 2008).

Two different approaches to performing activity control are introduced that would suit various levels of physical as well as cognitive impairments of the patient. One approach was based on adaptation of treadmill speed during walking; the other was based on instructions given by visual information from a virtual environment. These two methods were experimentally evaluated using the Lokomat gait orthosis (Colombo et al., 2000; Riener et al., 2010) in three experiments with five stroke patients each.

2.2.1 Definition of Physical Effort

The robot could be operated with varying degrees of supportive force, which significantly influenced patient participation. If the impedance controller was set stiff, the robot was position controlled. If the impedance was set low, the patient could lead the walking movement him or herself. At high assistive forces, the patient was able to push against the orthosis in direction of the walking movement, thereby overemphasizing the walking movement. Conversely, the patient could also behave passively and obtain a major contribution of the torques required for walking from the robot. The lower the impedance of the robot, the more torque the patient had to generate him or herself. At zero impedance, the robot did not provide any torque to assist the movement but behaved transparently by hiding its gravitational, Coriolis and friction forces, as well as its inertia.

I defined patient activity during robot-assisted gait rehabilitation by means of WIT to be high when the patient actively contributed to the walking movement.

The patient had to keep the assistive torque of the gait orthosis to a minimum and would perform the walking movement him or herself. At high impedance, the walking movement was fully prescribed by the gait robot. The patient was then able to perform active voluntary movements; pushing into the orthosis, the patient could overemphasize the walking movement and expended additional energy.

Conversely, patient activity was defined as low if the patient did not actively contribute to the walking movement and WIT values were negative. This was also only possible at high impedance settings, as the gait robot then provided most of the torque necessary to perform the walking movement and the patient was mostly moved by the gait robot in the walking trajectory. Gait speed, amount of body weight support and amount of assistive force generated by the orthoses influenced the effort necessary to perform the walking movement. The patient was forced to expend more energy during training when gait speed was increased, body weight support decreased or assistive force was decreased (Thomas et al., 2007).

2.2.2 Quantifying Physical Effort

Physiological Quantification

Quantifying physical effort via a physiological parameter, HR measurements were used which are known to reflected patient participation during body weight supported treadmill training (Thomas et al., 2007). The electrocardiogram (ECG) was measured with three surface electrodes. One electrode was affixed 2 cm below the right clavicula between the 1st and the 2nd rib, one was affixed at the 5th intercostal space on the mid axillary line on the left side of the body, and a ground electrode was affixed to the right acromion. The ECG was recorded with a gTec (www.gtec.at) amplifier, sampled at 512 Hz, filtered with a 50 Hz notch filter and bandpassed with a 20-50 Hz Butterworth filter of 4th order. HR was then extracted in real time using a custom steep slope detection algorithm adapted from (Malik, 1996). All software was implemented in Matlab 2008b (www.mathworks.com). While changes in HR have a clear dependence to changes in physical effort, other stimuli such as negative emotions or stress effect HR as well. Please refer to the Discussion of this chapter for detailed explanations.

Biomechanical Quantification

To quantify physical effort from a biomechanical measure, the WIT between robot and patient were computed, recorded from hip and knee joints of both legs, using the standard Lokomat force sensors located in line with the linear guides.

For each step, the interaction torques of all four joints were computed from the force recordings, weighted using the weighting function of Banz et al. (Banz et al., 2008) and summed up. In previous work, Banz et al. 2008 had investigated whether the interaction torques could be used to distinguish a physiologically desired movement pattern that would be beneficial for rehabilitation outcome from a walking pattern that would not be desired, as rated by expert physiotherapists. The result was the weighting function that is used to compute the WIT values (Banz et al., 2008) . The WIT has a high positive value if the patient performs an active movement which is therapeutically desired and a negative value if the patient is passive or resists the walking pattern of the orthosis. Values around zero mean that the patient is able to minimize the interaction torques between his legs and the orthosis. Details on the computation and their physiotherapeutic interpretation can be found in (Lunenburger et al., 2004, 2007; Banz et al., 2008).

The raw, unprocessed interaction torques between the patient and the orthosis were not used to quantify, how much the patient contributed to the walking movement him or herself. Raw torque exchange is not a suitable measure for patient activity, as therapeutically undesired movements can result in large interaction torques between Lokomat and human. Spasticity, for example, can cause large interaction torques, but usually does not contribute to a physiologically meaningful gait pattern.

2.2.3 Controlling Physical Effort with Visual Instructions

Patients that were cognitively capable of understanding virtual task and producing voluntary force were provided with real time feedback on their current activity using visual displays. With voluntary physical pushing effort, the patient had to match the current effort to a desired effort displayed on a screen. In this case, the control loop was closed via a visual feedback loop, as the instructions to the patient were given visually. The virtual stimulus was designed to be as easy and intuitive as possible such that patients with cognitive impairments were able to understand and perform the task. All action in the virtual environment took place on a straight path in the middle of the screen such that patients that suffered from neglect syndrome in the visual field could use the virtual environment.

The desired patient effort, quantified either by HR or by WIT, was displayed by the position of a dog walking in a virtual forest scenario (Fig 2.1). The current patient effort was displayed as a white circular area on the floor of the virtual scenario. The error between desired and recorded activity was mapped with a P gain to a distance between the white circle and the dog (Fig. 2.2). By increasing effort, the white circle moved faster, by decreasing effort, the white circle moved forward slower. The patients were instructed to place the white circle underneath the dog. This means the patients knew they had to increase their effort if the dog was walking too far ahead of the white circle and decrease their effort if the dog was walking behind the white circle. The distance between dog and the circle, therefore, displayed the difference between the desired and the actual effort of the

2.2 Methods

Figure 2.1: Virtual scenario used for control of WIT and HR. The distance between the dog (desired effort) and the white circle (actual effort) is the visual instruction to the patient. By increasing or decreasing his/her effort, the patient controlled the walking speed of the white circle

patient. Control of HR and WIT via visual stimuli was performed with the same stimulus for both measures of patient activity.

2.2.4 Controlling Heart Rate Using Treadmill Speed

Adaptation of treadmill speed allowed us to control patient activity to a desired temporal profile without the use of a virtual task. This would be necessary when the patient is cognitively not capable of understanding visual feedback, or physically not capable of exerting enough voluntary physical effort to control the virtual task. A higher physical load was imposed on the patient by increasing gait speed such that the patient was forced into a walking movement, which required increased activity. Conversely, lower gait speeds demanded less physical activity of the patient.

HR was controlled using a PI controller with anti-windup that adapted treadmill speed (Fig. 2.3). PI control was chosen as it is well established in control systems design and has previously been used in HR control of healthy subjects. A discussion of advantages and disadvantages of previous approaches to HR control and their applicability in stroke subjects can be found in the Discussion section.

P and I controller gains were set to 0.05 and 0.01, respectively. The gains were tuned in pre-experiments using the Ziegler-Nichols method (Guzella, 2007), a stan-

2 Controlling Physiological States during Robot-Assisted Gait Training

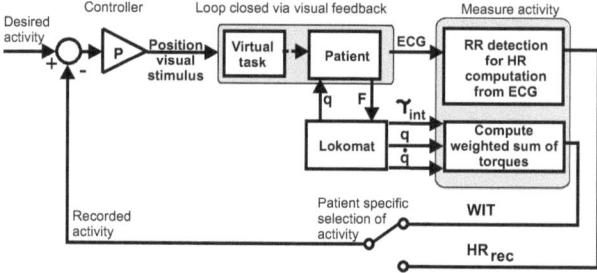

Figure 2.2: Control scheme for control of activity via a visual stimulus. Active participation is measured by HR or weighted interaction torques (WIT). The control loop is closed by the visual feedback to the patient. τ_{int} are the interaction torques between Lokomat and human. If HR control is chosen, mean HR is extracted in real time from the ECG and compared to a desired HR value. If WIT is controlled to a desired value, the current WIT values are computed from angular position (q), angular velocity ($\dot{q}$) and the interaction torques. The position of the visual stimulus is computed with a P gain.

dard methods for controller gain tuning when recorded data of a step input is available, and then fixed for all other subjects. Baseline HR was recorded at 1.5 km/h.

2.2.5 Experimental Protocols

Three experiments were performed (Tab. 2.1) to control HR via treadmill speed (experiment 1), via a visual stimulus (experiment 2) and WIT via the virtual task (experiment 3). Control of WIT was only performed using a virtual task, not via adaptation of treadmill speed. As described in the section "Quantifying patient participation", the patient had to actively participate in the walking movement to reach a high WIT value. The virtual task gave direct feedback on the current WIT and the patient could react to this visual feedback by increasing his or her activity voluntarily. Same was true for HR control via visual feedback. Treadmill speed as a control variable for HR was also used, as increased walking speed led to increased energy expenditure and, therefore, to increased HR. High WIT values, however, required the subject to voluntarily perform the walking movement in a therapeutically desired way, which was not controllable by treadmill speed alone.

All three experiments were performed with 5 stroke patients each, resulting in recordings of 15 patients (Tab. 2.2). Approval for all studies was obtained from local ethics committees, and all subjects or their legal representative gave written

2.2 Methods

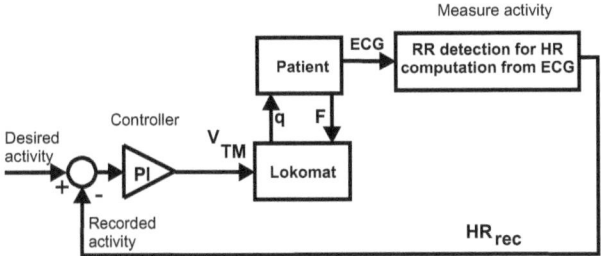

Figure 2.3: Control scheme for PI control of HR via treadmill speed. q are the orthosis joint angles, F are the forces exchanged between Lokomat and human. Mean HR is extracted in real time from the ECG and compared to a desired HR value. The error is fed into a PI controller that sets the gait speed of treadmill and Lokomat.

	Patient participation quantified by	
Control via	HR	WIT
Treadmill speed	Experiment 1	-
Visual Stimulus	Experiment 2	Experiment 3

Table 2.1: Overview over the experiments performed. Each experiment was performed with 5 stroke subjects

informed consent before data collection. The gait orthosis Lokomat (Colombo et al., 2000; Riener et al., 2010) (Hocoma Inc., Volketswil, www.hocoma.com) was used for all experiments, but the proposed approaches are generalizable to any gait robot that is equipped with force sensors. In the experiments with virtual environments, subjects walked in the Lokomat at 2 km/h with maximal supportive force by the robot and individual body weight support settings determined by the therapist.

During HR control via treadmill speed, the combination of the path control mode (Duschau-Wicke et al., 2010) with a modified Lokomat software allowed walking speeds up to 4 km/h. The maximal walking speed was determined for each patient, as not all patients were physically capable to walk at the maximal possible gait speed of 4 km/h. Minimum body weight support was identified for each patient individually by decreasing unloading at maximal walking speed in steps of 1 kg. Minimum body weight support was set right before the gait pattern degraded visibly as rated by the attending physiotherapist. The unloading was then kept constant over the whole training session. All patients of HR control experiments were instructed to refrain from coffee, nicotine, chocolate, black tea and energy drinks up to 4 hours prior to the experiment. HR control was only performed with patients that did not take beta blocking medication. All patients or their legal representative gave informed consent.

In all experiments, subjects were allowed to walk for ten minutes to get acquainted to the Lokomat. During these ten minutes, baseline HR was determined at a gait speed of 1.5 km/h. Lower gait speeds were reported to feel unnatural by the patients. If patients were walking in a virtual environment, they could also exercise the task during these ten minutes. HR or WIT were controlled to a desired temporal profile which included four distinct conditions of patient activity: low, intermediate, high and very high (100%, 33%, 66%, 100% as shown in Fig. 2.4, Fig. 2.5 and Fig. 2.6, dashed line). Each condition was set to be three minutes long. Three minutes was a tradeoff between reaching steady state of HR and keeping the duration of the experiment sufficiently short such that the whole recording was kept below 30 minutes, which was requested by therapist, thus avoiding overexertion of the patient.

The desired profile was scaled in amplitude to the maximal and minimal values of HR and WIT of each subject individually. In the virtual reality approach, patient specific limits of HR or WIT were identified during the exercise time, by asking the subjects to perform at their respective maximal and minimal level of activity. In the treadmill speed approach, the maximal HR was identified before the experiment by letting the patient walk at his/her maximally tolerable walking speed.

2.2.6 Controller Performance Evaluation

Controller performance was evaluated by normalizing the recorded HR / WIT for each patient after his or her minimal and maximal HR / WIT. Data was then low pass filtered with a 4th order, zero-phase Butterworth filter with cut-off frequency of 1 Hz to show the underlying trend. For HR data, the cut-off frequency of 1 Hz was experimentally determined to remove HR fluctuations caused by HR variability. Mean and standard error of HR and WIT were computed, taken over the last minute of each condition to quantify steady state behavior rather than transient behavior. Statistical tests were used to compare the four desired conditions of physical effort (dashed lines in Fig. 2.4, Fig. 2.5 and Fig. 2.6). In addition, the three approaches were compared amongst each other to investigate, if the results of the three different approaches differed significantly from each other. Both tests were done with a Friedman test with Bonferroni correction. Significance level was set to 0.05 for all tests. Data processing was done using Matlab (www.mathworks.com), statistical analysis was performed using IBM SPSS (www.spss.com).

2.2 Methods

Pat no.	Gender	Age [y]	Time since incident [m]	Lesion	β blocker	FAC	Cognitive deficits	
1	m	43	29	r. hemorrhagic	no	3	attention deficit	HR control via treadmill speed
2	w	52	5	l. ischemic	no	3	attention deficit	
3	w	33	22	l. ischemic	no	2	n.a.	
4	m	49	29	l. hemorrhagic	no	2	attention deficit	
5	m	57	23	r. ischemic	no	2	n.a.	
6	m	36	5	r. ischemic	no	1	Small memory deficits	HR control via virtual stimuli
7	m	71	2.5	r. ischemic	no	2	Neglect left	
8	m	68	2.5	r. ischemic	no	4	none	
9	m	55	3.5	r. hemorrhagic	no	0	Small attention deficit, neglect left	
10	m	67	2	r. ischemic	no	3	Medium attention deficit, neglect left	
11	m	65	1.5	l. ischemic	yes	5	Small attention deficits	WIT control via virtual stimuli
12	m	62	2	r. ischemic	yes	n.a.	None	
13	m	68	2.5	r. ischemic	no	4	None	
14	m	67	2	r. ischemic	no	3	Small attention deficit	
15	f	58	3.5	r. ischemic	no	n.a.	Small attention deficit	

Table 2.2: Characteristics of patients of HR and WIT control experiments. Gender: m=male, f=female, l=left, r=right, FAC=Functional Ambulation Classification (0= patient can only walk with the help of at least 2 people, 5= patient is a communal walker), n.a.= data not available

2.3 Results

2.3.1 Control of HR via Treadmill Speed

HR control of stroke patients via adaption of treadmill speed was performed successfully (Fig. 2.4). Minimal and maximal HR values were used for normalization (summarized in Tab. 2.3) such that the average tracking performance of the controller could be computed. The mean HR values of the last minute of each condition were computed and summarized on the top of Fig. 2.4. Patient 2 had to be excluded from the analysis, as she could not complete the desired protocol due to spasticity in the ankle joint of the affected leg caused by the physical effort of walking on the treadmill. All of the other subjects informally reported to be very exhausted at the end of the recording.

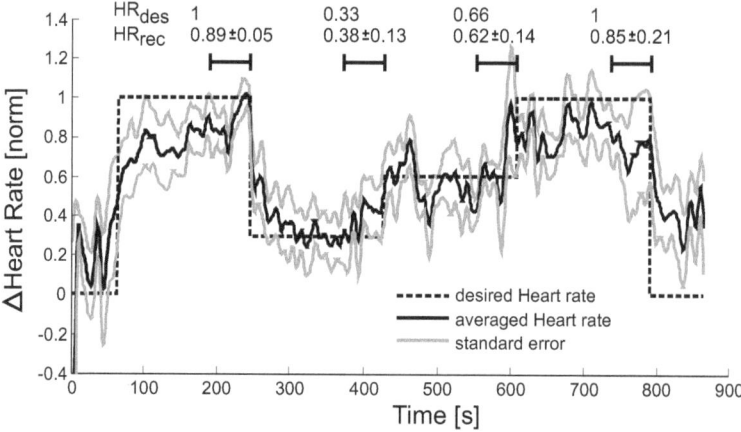

Figure 2.4: Results of PI control of HR via adaptation of treadmill speed. Results were normalized and filtered with a low pass of 0.5 Hz to show the underlying trend.

2.3.2 Control of HR via Visual Instructions

HR control of stroke patients was successfully performed via visual instructions from a virtual environment. As described above, subjects obtained instruction from the virtual environment to increase or decrease their voluntary physical effort and thereby their HR. The normalized results of HR control via visual stimuli are depicted in Fig. 2.5. The success of controlling HR with visual instructions was

2.3 Results

Patient	Minimum HR	Maximum HR
1	60	75
2	97	107
3	79	89
4	80	92
5	85	97

Table 2.3: Minimum and maximum HR values of each patient used for control of HR via treadmill speed. These values were determined during the initial baseline recording and used for normalization during data processing. The desired HR profile to be tracked was scaled with these values to enable patient-specific exercise.

quantified by mean HR and standard error of all five subjects. The mean HR values of the last minute of each condition were computed and summarized on the top of Fig. 2.5. It was necessary to adjust the baseline and maximal HR increase individually for each subject to provide patient-specific control of HR. In average, it was possible to increase HR by 11 ± 4 beats per minute. Normalization values are summarized Tab. 2.4.

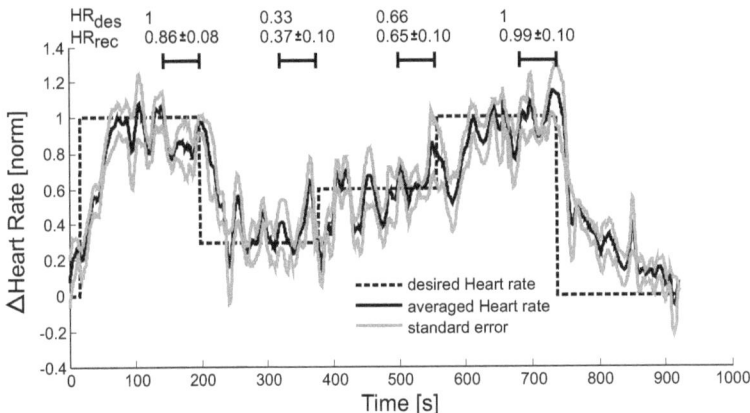

Figure 2.5: Results of HR control via visual instructions. Results were normalized and filtered with a low pass of 0.5 Hz to show the underlying trend.

2 Controlling Physiological States during Robot-Assisted Gait Training

Patient	Minimum HR	Maximum HR
1	90	105
2	75	90
3	110	117
4	93	103
5	104	112

Table 2.4: Minimum and maximum HR values of each patient used for control of HR via visual instructions. These values were used determined during the initial baseline recording and used for normalization during data processing. The desired HR profile to be tracked was scaled with these values to enable patient-specific exercise.

2.3.3 Control of WIT via Visual Stimuli

Control of WIT by means of a virtual stimulus was also performed successfully in five stroke patients (Fig. 2.6). Tracking performance was quantified by mean WIT and standard error of all five subjects. The mean WIT values of the last minute of each condition were computed and summarized on the top of Fig. 2.6. It was necessary to adjust the baseline and maximal WIT increase individually for each subject to provide patient-specific control of WIT. Normalization values are summarized in Tab. 2.5.

While the levels of 30% and 60% of maximal WIT could be tracked well, subjects had problems in reaching desired maximal WIT. They could reach the desired maximal level for short time, but got exhausted too fast to keep the effort at this level.

Patient	Minimum WIT	Maximum WIT
1	0	90
2	-90	27
3	-27	81
4	-36	9
5	-45	0

Table 2.5: Minimum and maximum WIT values of each patient used for control of WIT via visual instructions. These values were determined during the initial baseline recording and used for normalization during data processing. The desired WIT profile to be tracked was scaled with these values to enable patient-specific exercise.

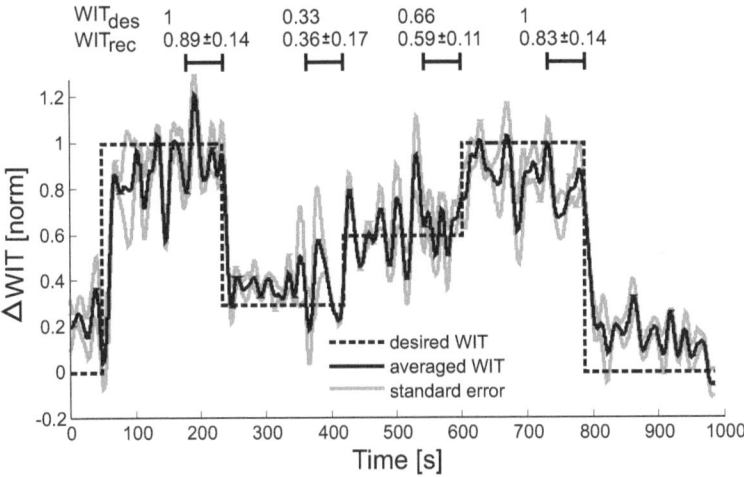

Figure 2.6: Results of WIT control via visual instructions. Data was normalized and filtered with a 1 Hz lowpass to show underlying trend.

2.3.4 Statistical Comparison Between the Three Approaches

The statistical analysis of each control approach showed that subjects could track the desired performance condition A-D (100%, 33%, 66%, 100% as shown in Fig. 2.4, dashed line) with all three approaches (Fig. 2.7, top). The comparison between the three different approaches showed that all approaches worked equally well for all conditions A-D (Fig. 2.7, bottom). The Friedman test did not result in significant differences between the different approaches.

2 Controlling Physiological States during Robot-Assisted Gait Training

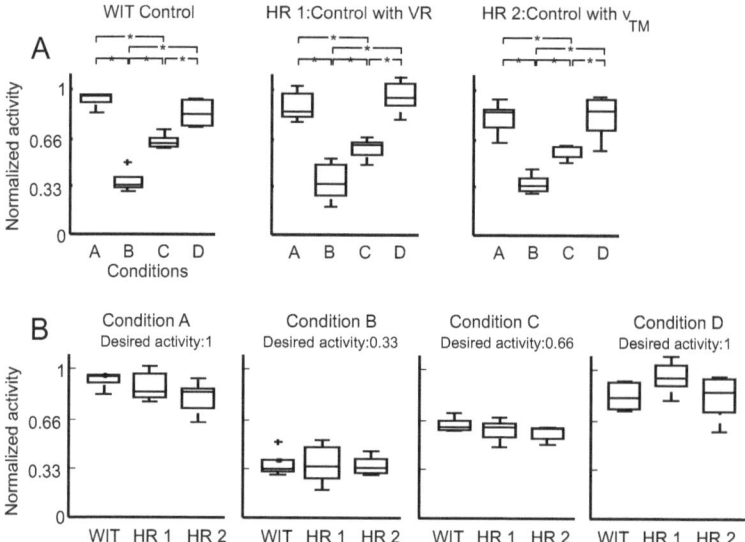

Figure 2.7: Boxplots comparing the three different approaches. WIT = WIT control with VR. HR1 = HR control with VR. HR2 = HR control with treadmill speed. Conditions A, B, C and D refer to the different levels of activity (100%, 33%, 66%, 100%). A: within one control approach, all conditions (except A compared to D) differ statistically. B: No significant differences were found between WIT, HR1 and HR2 for any condition.

2.4 Discussion

The overall goal of this work was to investigate approaches to controlling active participation in stroke patients during robot-assisted gait therapy. Patient effort was quantified in two ways: by HR and by a weighted sum of interaction torques (WIT). For validation of the proposed approaches, three experiments were performed, in which stroke patient's HR and WIT were controlled to a desired temporal profile.

Although active physical participation during gait rehabilitation was shown to be crucial for recovery from stroke (Lotze et al., 2003), patients can behave passively during rehabilitation and might therefore not maximally benefit from the gait training. This might explain why several studies reported inconclusive results on the effects of robot-assisted gait therapy compared to manually-assisted gait therapy after stroke or spinal cord injury (Mehrholz et al., 2008, 2007).

Patient participation was successfully controlled to a desired level (Fig. 2.4, Fig. 2.5 and Fig. 2.6). Depending on the patient's cognitive capabilities, this was either done by voluntary patient effort using visual instructions or by forcing the patient to varying physical effort by adapting the treadmill speed. In addition to adapting to the cognitive capacities of the patient, an initial magnitude scaling of the desired temporal control profile allowed adaptation to patient individual physical capabilities. Four levels of patient activity were targeted: 100%, 66%, 33% and again 100% of maximal participation (Fig. 2.4, dashed line). Using three different approaches, all patients could equally well track the desired temporal profile, independent of their cognitive or motor impairments (Fig. 2.7, top). Results of the Friedman test showed no statistical differences in their applicability to patients (Fig. 2.7, bottom). This framework is intended to enable therapists to challenge the patient to active participation by automatically controlling the patient effort to a desired level.

2.4.1 Patient Individual Control of Participation

One problem of controlling patient participation is the necessity of scaling the desired participation to a level, where the patient is able to perform at his or her individual capabilities.

The maximal and minimal WIT values (Tab. 2.5) reflect the individual physical ability of each patient. Although the WIT is a unit less quantity, healthy subjects could reach values between -400 while strongly resisting the orthosis movement during the walking pattern to +400 while maximally overemphasizing the walking movement and pushing into the orthosis. Patient 5 could only reach a maximal value of 0 which corresponds to the ability to perform the walking movement himself without being able to generate any additional pushing force in walking direction. Patient 1 on the other hand could reach a value of 90 which means that this patient was able to voluntarily push into the orthosis. Nevertheless, both patients could receive a challenging training that was adjusted to their individual capabilities.

During control of HR in experiment 1 and 2, the HR recorded at baseline reached from 60 bpm (Tab. 2.2, patient 1) to 110 bpm (Tab. 2.2, patient 8). With the limitations of gait speed imposed by the Lokomat and the physical abilities of the patients, some patients could only reach an increase in HR of 7 bpm (Tab. 2.2, patient 8), while others could be controlled in a range of 15 bpm. Challenging training sessions could still be provided to the patients, independent on their individual physical capabilities, as all patients informally reported to be exhausted after HR control experiments,

2.4.2 A Metric for Patient Individual Control of Physical Activity

Based on the three successful approaches to controlling patient participation, I propose a metric which enables clinicians to select the best strategy for each patient, according to the patient's physical and cognitive capabilities. Controlling WIT requires the patient to have a cognitive understanding of a therapeutically desired gait pattern and the physical capability to alter the current gait pattern according to the performance feedback. I therefore consider WIT control to be the most challenging task that patients can perform.

HR, however, will increase, as soon as the patient produces voluntary force against the position controlled orthosis, regardless if the movement is therapeutically beneficial or not. Resisting the movement of the orthosis would not be therapeutically desired, but the effort involved with the muscle contractions would increase HR. In order to control a virtual task with his/her HR, the patient needs to have the cognitive understanding of the task and must be able to produce physical effort. As this physical effort does not require the capability of the patient to adapt his or her gait pattern to a therapeutically desired pattern, the physical as well as the cognitive abilities of the patient do not have to be as intact as during WIT control. Patients with severe impairments might not be able to understand visual performance feedback or not be capable of generating enough pushing effort to increase their HR. For these patients, I propose HR control via adaptation of treadmill speed since higher gait speed in the Lokomat will increase HR regardless of the ability of the patient to voluntarily push into the orthosis.

While this metric will allow controlling participation in a wide range of patient groups, not all patient groups will benefit from it. Patients taking Beta blocking medication will not be able to exercise in HR control mode, as Beta blockers were shown to decrease HR variability and limit the adaptation of HR to physical stress (Cook et al., 1991). These patients can still benefit from WIT control. Patients that are unable to produce directed, voluntary effort will neither be able to increase

2.4 Discussion

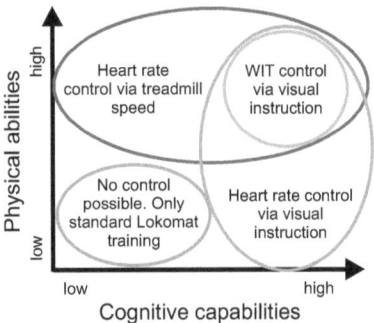

Figure 2.8: Selection matrix for optimal training of stroke patients, depending on the patient's cognitive and physical impairments. Mildly affected patients can exercise in all three modes: HR control via treadmill speed and visual instructions and WIT (Banz et al., 2008) control via visual instructions. Patients with strong cognitive deficits might only be able to exercise in HR control mode via treadmill speed. Patients that are capable of understanding a virtual task but are physically limited in their capabilities of controlling the WIT to a desired value can still exercise in HR control mode via visual instructions.

their HR by increased power expenditure, nor will they be able to control their WIT to a desired level. Furthermore, if the cognitive impairment does not allow the use of visual instructions and physical impairment prohibit walking at treadmill speeds that allow HR control via adaptation of treadmill speed, then control of patient participation will not be possible.

During the study design, 15 different subjects were recruited that had never experienced virtual reality feedback in the Lokomat for the three experiments. This broad patient base allowed to investigate whether the proposed methods would worked with a variety of different impairments. The next step will be a larger study that can provide statistical evidence for the metric proposed in Fig. 2.8. An objective rating for the cognitive impairments of subjects such as the mini-mental state estimation (Folstein et al., 1975) will then also be collected.

2.4.3 Cardiovascular Training After Stroke

Our proposed method combines the advantages of virtual reality augmented gait training with the benefits of cardiovascular training. Non-ambulatory patients that use HR control during Lokomat walking are able to combine gait training with cardiovascular training. The benefits of cardiovascular training come at no extra cost to benefits of gait rehabilitation.

The use of virtual reality might increase the training efficacy of robot assisted gait therapy compared to training without virtual environments, as recently demonstrated by studies of Mirelman (Mirelman et al., 2009) and Brütsch (Brütsch et al., 2010). HR control via visual feedback has not been performed during robot assisted gait training before. However, oxygen uptake was controlled to a desired trajectory via volitional pushing effort during robot assisted gait training (Pennycott et al., 2010). Subjects had to increase and decrease their effort (and thereby their energy expenditure) according to a visual display which coded the deviation from a desired oxygen uptake value.

Cardiovascular training, such as treadmill based HR control, was shown to be beneficial to stroke survivors during gait rehabilitation (Gordon et al., 2004). Depending on the degree of impairments caused by the lesion, this training has been performed either on treadmills for less severe cases or on stationary bicycles in severely affected patients. Particularly non-ambulatory patients were not able to exercise on treadmills, but had to use stationary bicycles instead, where the problems of coordination and balance during walking did not need to be taken into consideration.

To my best knowledge, there has been no study in which HR of stroke patients was controlled during treadmill walking. In healthy subjects, treadmill based HR control has been successfully demonstrated using PID or H_{inf} control (Su et al., 2007a,b; Cheng et al., 2008). In these studies, HR increases of 30 beats per minute (bpm) were demonstrated; it was only possible to reach an average HR increase of 12 bpm using treadmill speed as control signal. This seems to be a very small increase compared to the results obtained in healthy subjects. However, previous approaches to HR control of healthy subjects were performed at walking speeds starting at 3.6 km/h (Su et al., 2007a,b; Cheng et al., 2008), which are not feasible for most patients. Within the patient group, only one individual was able to walk at speeds higher than 3.6 km/h. Pennycott et al. (Pennycott et al., 2010) controlled oxygen uptake during Lokomat walking, however only in healthy subjects and with the drawback, that the method needed an initialization time for parameter identification, which would shorten the duration available for actual cardiovascular training in patients.

In addition to treadmill speed, the amount of body weight support would have an impact on the effort which patients have to expend during walking. Unloading was shown to alter HR at constant walking speeds (Thomas et al., 2007). In accordance with the therapists, it was decided not to use body weight support as a control variable. Increased body weight support reduced the loading to be carried by the patient during gait. High loading of the patient during treadmill training was shown to be a key factor for rehabilitation success (Dietz and Duysens, 2000).

In order to maximize the quality of gait training, it was decided to set body weight support to a fixed, patient-specific minimal value.

2.4.4 Clinical Applicability of Patient Activity Control

Heart Rate

Despite all the advantages of HR control, there are major drawbacks compared to WIT control. First, patients have to refrain from consuming any substance that might influence HR such as beta blockers, coffee, nicotine or tea. Controlling patient behavior in this way might not be possible in a clinical setting. In addition, HR control requires an additional computer for recording of ECG, and additional time for attachment of the electrodes before the training. Also, carefully monitoring of the ECG signal quality is necessary, which might degrade over the time course of a training session due to sweating or artifacts induced by the body weight support harness. Finally, as described in chapter 1.5.5, HR does not only increase with increasing physical activity, but was also found to increase due to stress or negative emotions (Levenson et al., 1990; Andreassi, 2007) and decreased in reaction to unpleasant stimuli (Palomba et al., 2000; Gomez et al., 2005; Codispoti et al., 2008). It is therefore possible that changes in HR recorded during the training are not exclusively caused by cardiovascular training, but by external and internal factors that are difficult to quantify.

WIT

WIT in comparison does not require additional setup time, no additional sensors and will also work in patients that are under the influence of any of the above mentioned substances. There is one drawback to quantifying patient effort via WIT. While HR was shown to increase with increasing effort and energy consumption (Thomas et al., 2007), WIT does not necessarily reflect energy consumption. A patient could theoretically have a high value of co-contraction, independent on his or her WIT values. This increased metabolic energy demand would be reflected in HR, but not in WIT. However, at this time, WIT control seems to be more likely to find transfer into a standard clinical setting with patients that are cognitively capable of understanding a virtual task.

2.5 Contribution

The methods presented in this chapter were developed as part of my thesis. The patient data for effort control via virtual environments was recorded by my colleagues Jeannine Bergmann and Carmen Krewer from the 'Neurologische Klinik

Bad Aibling'. The final versions of the virtual environment was programmed by Lukas Zimmerli.

3 Controlling Psychological States during Robot-Assisted Gait Training

3.1 Introduction

Active cognitive engagement in the rehabilitation process, which was shown to be a key factor for successful rehabilitation (Maclean and Pound, 2000a), cannot be assessed easily and is therefore often neglected. Cognitively challenging training sessions were shown to be key requirements for the success of motor learning in general and in rehabilitation (Maclean and Pound, 2000a; Lotze et al., 2003; Kaelin-Lang et al., 2005). In addition, research in healthy subjects suggests that motor learning decreases in the presence of a distracting cognitive task, which presents a cognitively over-challenging situation (Redding et al., 1992; Taylor and Thoroughman, 2008). From motor learning theory it is known that the learning rate is maximal at a task difficulty level that positively challenges and excites subjects while not being too stressful or boring (Guadagnoli and Lee, 2004). A task which is too easy for the subject will be perceived as boring, a task which is too difficult will overstress the subject, while an optimally challenging task should induce maximal cognitive engagement and optimal physical participation.

Quantifying and controlling cognitive load in neuro-rehabilitation to avoid tasks that are cognitively too demanding or too easy, has the potential to increase motor learning and thereby the training efficiency and therapeutic outcome of neurological rehabilitation (Maclean and Pound, 2000a; Kaelin-Lang et al., 2005).

Despite existing tools used to modulate patient motivation such as virtual environments (Holden, 2005), the rehabilitation environment does not yet adapt to the cognitive load of the patient. One major reason is that the current cognitive load of patients cannot be objectively assessed. Questionnaires can be used to obtain subjective information, but only at discrete time-points after training has ceased. In the present state of the art, cognitive engagement of subjects is for example quantified via the "Intrinsic Motivation Inventory" (McAuley et al., 1989). Also, during gait rehabilitation, questionnaires are not appropriate for continuous, objective assessment of the psychological state of the patient. In addition, neurological patients with severe cognitive deficits or aphasia might not be able to understand and respond appropriately to the questions.

As already detailed in chapter 1, physiological signals used to quantify the psychological state of a subject can be influenced not only by cognitive and physical effort, but also by unpleasant stimuli or fear (Levenson et al., 1990; Andreassi, 2007; Palomba et al., 2000; Gomez et al., 2005; Codispoti et al., 2008; Ohsuga et al., 2001; Brown and Schwartz, 1980; Larsen et al., 2003). To exclude the possibility that changes in physiology recorded during virtual task experiments were caused by emotions such as fear of the training or any other unpleasant stimulus, it was necessary to quantify the influence of such stimuli. Measurements of heart rate (HR), breathing frequency, skin conductance and skin temperature were used to investigate, how unpleasant stimuli of fear would find reflection in psychophysiological recordings. Experiments were therefore performed, in which the physiological reaction to an unpleasant or scary virtual stimulus was recorded. For ethical reasons, these experiments were only conducted with healthy subjects.

The next goal was to investigate, if virtual environments could induce different levels of cognitive load and the potential of physiological signals to identify the current cognitive state by using descriptive statistics. Subjects were presented with a virtual task which was used to induce three different levels of cognitive engagement. Again, measurements of HR, breathing frequency, skin conductance and skin temperature were used to investigate, how different levels of cognitive load would find reflection in psychophysiological recordings. Attaching sensors for physiological recordings on the subject's body was time and labor intensive requiring attention of the therapist and reducing the time a patient could exercise in the Lokomat. To improve clinical applicability of the approach, I investigated which physiological signals contained most of the information and which sensors might not be necessary for future applications. Principal component analysis (PCA) allowed identification of the signals that explained most of the variance in the data.

Then, three different classifiers were trained to investigate the possibility of identifying the current state of cognitive engagement directly from physiology. The classifiers were trained with data from open loop experiments in healthy subjects and stroke patients. A Neural Network, which was trained with all physiological signals and with a reduced dataset of signals, which were shown to be dominant in the PCA. A classic Linear Discriminant Analysis Classifier (LDA), and a Kalman adaptive LDA.

In the last step, bio-cooperative closed loop control of cognitive load was performed in both healthy subjects and patients. Closed coop control was only performed using the LDA and the KALDA system, as the Neural Network was found unpractical, as it required recording training data at the beginning of each session. Performance metrics were investigated as a proxy for psychophysiological signals. While performance metrics might be less accurate, they are more practical to obtain in a clinical setting. I therefore compared a closed loop controller using physiological signals with a simple task performance controller that only controlled task success without the use of physiological signals.

The methods and results presented in this chapter have been previously published in two journal papers (Koenig et al., 2011a,e) and two conference papers (Koenig et al., 2010, 2011c). The text used from (Koenig et al., 2011a) and (Koenig et al., 2011c) is ©2011 IEEE, both reprinted, with permission. The text used from (Koenig et al., 2011e) first appeared in the Journal of Rehabilitation Research and Development and is reprinted with permission.

3.2 Methods

3.2.1 Common Methods

Some methods were common to all goals of this chapter: the questionnaires, the attachment of the sensors for recordings of physiological signals and the experimental setup. The virtual task was always a collect and avoid task. However, two different implementations were used during different stages of the project, which is why the different tasks are explained independently.

Questionnaires

Three different levels of cognitive engagement were defined according to the circumplex model of affect (Russell, 1980) (Fig. 3.1), in which emotions are defined by two dimensions: valence (ranging from unpleasant to pleasant) and arousal (ranging from deactivation to activation). Virtual environments were used during robot assisted gait training to induce different levels of cognitive engagement in subjects. Challenging tasks in virtual environments were shown to have a positive, motivating effect during rehabilitation (Holden, 2005). In this context, boring, too stressful or optimally challenging tasks can be the result of a virtual task that is too easy, too difficult or appropriate for the patient's abilities. A virtual task which is too easy or under-challenging for the subject will be perceived as boring, a task which is too difficult will over-stress the subject, while an optimally challenging task should excite and motivate the subject and cause maximal cognitive engagement and optimal physical participation.

In addition, subjects were asked how difficult they perceived the task in terms of physical effort on a five point scale (Fig. 3.2). A score of 1 was not physically exhausting and 5 was extremely physical exhausting.

All questions were posed nonverbally as a pictorial questionnaire, as not to disturb the breathing frequency analysis by speaking and also to reduce the complexity of responding to the questionnaire for aphasic stroke patients or patients with cognitive impairments.

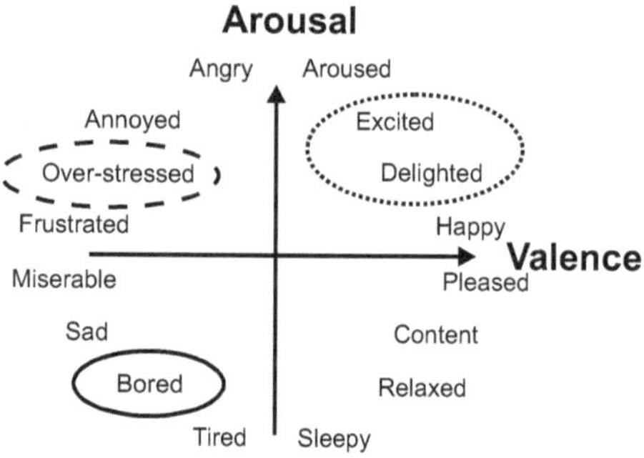

3 Controlling Psychological States during Robot-Assisted Gait Training

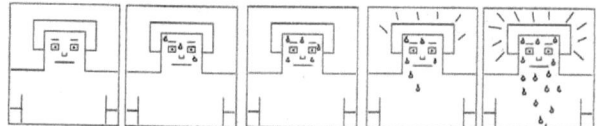

Figure 3.1: Three conditions were defined according to the circumplex model of affect. The circle with the solid line represents the under-challenged condition, the circle with the dotted line the challenged condition and the circle with the dashed line the over-challenged condition. (Adapted from Russell 1980, p. 1167)

Figure 3.2: Pictorial questionnaire on physical effort, ranging from 1 = not physically exhausting to 5 = extremely physical exhausting

Physiological Recordings

In addition to the subjective questionnaires, the effects of different levels of cognitive engagement were objectively quantified by recordings of ECG, breathing, skin conductance and skin temperature. The ECG was measured with 3 surface electrodes. One electrode was affixed 2 cm below the right clavicula between the 1st and the 2nd rib, one was affixed at the 5th intercostal space on the mid axillary line on the left side of the body, and a ground electrode was affixed to the right acromion. HR was computed from ECG using a real time R-wave detection algorithm (adapted from (Christov, 2004), Fig. 3.3), HRV was computed as a discrete time series of consecutive RR intervals. The square root of the mean squared differences of successive normal-to-normal intervals (RMSSD) was computed from RR intervals for analysis of HRV in the time domain. In the frequency domain, analysis of HRV was performed by windowing the last 300 seconds of HR data with a Hamming window and performing a Fast Fourier Transform. Then, according to the recommendations of Malik et al. (Malik, 1996), the quotient of low frequency components over high frequency components (LF/HF). The respective frequency bands can be found in (Malik, 1996).

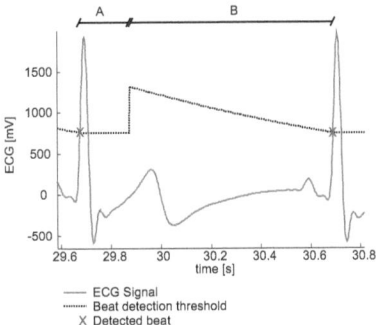

Figure 3.3: ECG signal and beat detection threshold when the ECG exceeds the threshold, a beat is detected. For 200ms after a detection, a beat is not allowed to be detected (1 - beat expectation between 200ms and 1200ms after the last beat). After 200ms, the threshold is increased to 0.6 of the maximum value of the last 5 beats. The amplitude is decreased over time (2 - steep slope threshold) until the next beat is detected.

Using a thermistor flow sensor placed underneath the nose, the breathing of subjects was recorded and breathing frequency was recorded using the same peak detection algorithm used in ECG processing. Changes in galvanic skin response were measured using two electrodes attached to the proximal phalanx of the second and the fourth fingers on the left hand or the unaffected hand in stroke patients. SCL was high-pass filtered with a fourth order, 20 Hz Butterworth filter to remove sensory noise. One SCR was detected from the skin conductance signal, when signal amplitude changed by at least 0.05 μSiemens in less than 5 s (Dawson et al., 2007). SCR were detected by fitting a polynomial through the last 30 seconds of data and detecting minima and maxima in the fitted polynomial. Skin temperature was measured on the distal phalanx of the fifth finger of the left hand or the unaffected hand in stroke patients and not processed further. Facial EMG was excluded, since some patients did not tolerate additional sensors attached to their faces while other patients suffered from half sided paralysis of their facial muscles. The signals were amplified with the g.USBamp of Guger Technologies, Graz, Austria (www.gtec.at). Signals were sampled at 512 Hz according to the recommendations of Malik (Malik, 1996). All signal processing software was written in Matlab 2008b (The Mathworks, Natick, MA, USA, www.mathworks.com).

Experimental Setup

The experimental setup consisted of three parts: a commercially available driven gait orthosis (DGO) commonly used in gait rehabilitation, the virtual reality display system and the measurement system for physiological signals Fig. 3.4. As DGO, the Lokomat (Hocoma Inc., www.hocoma.com) was used for the locomotion training. Drives on hip and knee joints provide torques to the subject and assist the locomotion on a treadmill by guiding the subject's legs along a predefined trajectory. Subjects were fixed into the DGO with a harness around the hip and cuffs around the legs. The feet of the subjects were passively lifted by elastic foot straps to prevent foot drop. Subjects were connected to a body weight support system. During walking the speed was kept constant at 2 km/h. Cadence had to be slightly adjusted to the individual leg length of the subjects. Subjects walked with the assistance of thirty percent body weight support. At Balgrist University hospital, the virtual environment was projected onto a 3x2m back-projection screen, which was mounted in front of the DGO with a Dolby 5.1 sound system for auditory feedback. At the Neurologische Klinik Bad Aibling, a 42 inch flatscreen TV with stereo sound was used. All physiological signals were recorded and amplified with the g.USBamp of Guger Technologies, Graz, Austria (www.gtec.at).

3.2.2 The Effects of Pleasant and Unpleasant Stimuli on Psychophysiological Recordings

Virtual Task

Three different virtual scenarios were presented to the subjects in randomized order. Scenario one was a 'pleasant' scenario in a forest with animals, nice sound and a sunset. Scenario two was unpleasant, as subjects walked in a 'rain' scenario with thunder and lightning on a narrow bridge over a canyon. Scenario three was a 'scary' scenario with fog, scary music and sounds, monsters and animals (Fig. 3.5).

Experimental Protocol

Each virtual scenario lasted for 2 min. At the beginning of the measurement, a 2.5 min baseline measurement without virtual reality was recorded. Between every virtual scenario a 2.5 min rest period took place which was also without any virtual reality. During the first 30 s of the baseline measurement and the rest periods subjects had to rate their level of arousal and valence by picking a figure (with a value from 1 to 5) from the SAM questionnaire. Heart rate, heart rate variability, breathing frequency, skin temperature, skin conductance responses and skin conductance level were recorded during the measurement. The measurement lasted approximately 14 min.

3.2 Methods

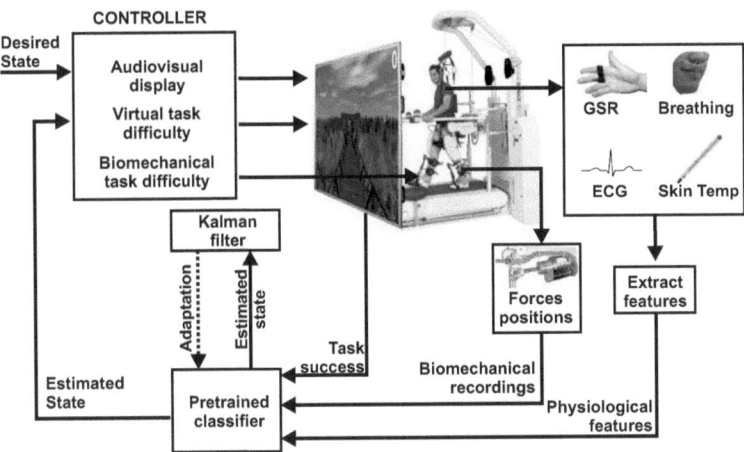

Figure 3.4: The multimodal controller setup for experiments on the effects of virtual environments on psychology. The Lokomat is equipped with a virtual reality system. On the screen, a virtual task with adaptable task difficulty is displayed. In addition to task difficulty, the atmosphere in the environment can be altered to additionally influence the psychological state of subjects. Physiological recordings, force data from the Lokomat and score information from the virtual environment are fed into a pre-trained, adaptive classifier. This classifier can update itself via a Kalman filter. The classifier estimates the current psychological state of the subject and outputs this estimate to the controller. The controller then adapts task difficulty and virtual atmosphere according to an internal set of expert rules. (©2010, 2011 IEEE)

Subjects

Five subjects were measured. Three female and two male with a mean age of 29.2 years (±5.3 years). All subjects gave written informed consent. As the exposure of patients to scary virtual scenarios was ethically not justifiable, only recordings with healthy subjects were performed.

37

3 Controlling Psychological States during Robot-Assisted Gait Training

Figure 3.5: Virtual environment with pleasant, unpleasant and scary stimuli. Left: pleasant and relaxing walk through a forest scenario. Middle: unpleasant walking over a swaying bridge. Right: scary scenario in a dark forest.

3.2.3 The Effect of Virtual Environments on Psychophysiological Recordings

Virtual Task

Subjects were provided with a virtual task which included two distinct actions at the same time, a biomechanical task and a cognitive task (Fig. 3.6). Subjects had to change walking direction in the virtual environment to collect items by walking into them (biomechanical task). To change the walking direction, subjects had to perform an active push-off in the terminal stance phase. To turn left, the subjects had to increase activity in the right leg during stance (Zimmerli et al., 2009). As the required physical effort to change walking direction was set individually, the challenge was to navigate through the virtual environment and collect items. In addition, subjects had to jump over barrels which rolled toward them by clicking a computer mouse button (cognitive task). Collected items added points to a counter, missed items and non-over jumped barrels subtracted points.

To create different task difficulty levels the distance between items were adjustable. Furthermore, the distance between the barrels and their speed were adjustable (Fig. 3.6). The task became more difficult by decreasing the distance between objects, the distance between barrels and by increasing the speed at which the barrels rolled towards the subject, as subjects had less time to react. As physical effort influenced the psycho-physiological recordings, the virtual reality (VR) task was chosen as a combination of coordination (change walking direction) and cognition (jump over the barrels). This allowed creating subject specific variation in task difficulty levels while keeping the physical effort necessary for successful task completion as low as possible.

Task difficulty was set individually by the experimenter for each healthy subject and patient during an initial practice time (Fig. 3.8). Task difficulty was adjusted as described in the previous section. In the under-challenged condition, the items

3.2 Methods

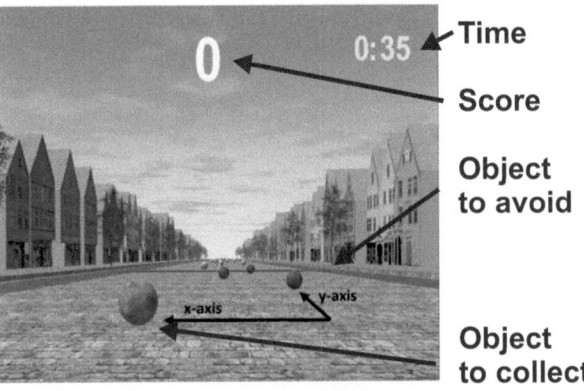

Figure 3.6: The virtual task with the x-axis representing the distance between the middle line and the items and the y-axis representing the distance between the items to be collected and the objects to be avoided. The x-and y-axis were adjustable to create different difficulty levels.

were placed such that the subjects could collect 100% of the items without major changes in walking direction. The challenged condition was defined by setting the distance between items and their distribution over the whole scenario such that subjects could collect 80-90% of objects and only missed approximately 10-20% of all items. In the over-challenged condition, the objects were distributed such that the subjects were left with less than 10% of the possible maximal score at the end of the condition (five minutes).

Inducing Different Levels of Cognitive Engagement

It was hypothesized that three different levels of cognitive engagement could be introduced during exercise by providing subjects with different levels of task level difficulty. The target cognitive engagement levels were: 1) a feeling of boredom; 2) a feeling of being motivated - excited, and 3) a feeling of being over-stressed (Fig. 3.7). Subjects were expected to be bored when the virtual task was so easy (under-challenged condition) and that no particular biomechanical or cognitive effort was necessary to successfully complete the task. Setting task level difficulty such that the task was difficult but feasible (challenged condition), subjects were expected to be motivated and excited. Setting the task level difficulty to be too difficult (over-challenged condition), subjects were expected to be over-stressed (Fig. 3.1). In the arousal-valence space (Fig. 3.1), the under-challenged condition would have a low level of arousal and a low level of valence, the challenged condition would have a high level of arousal and valence and the over-challenged condition would have a high level of arousal and a low level of valence.

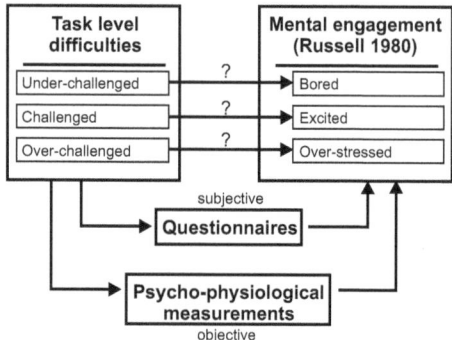

Figure 3.7: Inducing different states of cognitive engagement during gait training using the task level difficulty of a virtual reality task. Subjectively, questionnaires are used to establish the connection between the task level difficulty and the cognitive engagement. Objectively, psycho-physiological measurements are used.

The Self-Assessment Manikin (SAM) questionnaire was used to verify the hypothesis that the three conditions in the virtual task really resulted in a feeling of boredom, excitement and of being over-stressed (Fig. 3.1). The SAM is used to measure emotional response to different stimuli (Bradley and Lang, 1994), in particular the emotional responses arousal and valence. The dimension arousal ranged from "relaxed and sleepy" to "excited and extremely aroused". The dimension valence ranged from "unhappy" to "very happy" (Morris, 1995). Subjects were asked to respond to a 5-point scale by selecting the figure which best represented their current emotion. The value of 1 represented the lowest valence ("unhappy") and arousal ("sleepy") and 5 represented the highest valence ("very happy") and arousal ("excited"). After each task condition the subject was asked to respond to the SAM. Furthermore, subjects were asked to rate their physical effort on a scale from 1 to 5, where 1 was not physically exhausting and 5 was extremely physical exhausting. This nonverbal questionnaire was chosen so as not to disturb the breathing frequency analysis by speaking and also to give aphasic stroke patients or patients with cognitive impairments to reduce the complexity of responding to the questionnaire. All dimensions of the SAM questionnaire were tested statistically using the Friedman-test. A Wilcoxon test in combination with Bonferroni correction was performed for a paired comparison. The significance level was set at $p < 0.05$.

Evaluation of Physiological Recordings

Using descriptive statistics, it was investigated which physiological signals changed significantly between the different task level conditions. Only the last minute of each 5 minute condition was analyzed to ensure that steady state had been reached. Just like the questionnaire results, all conditions were tested using the Friedman test followed by a Wilcoxon test for paired comparison. Bonferroni correction corrected multiple errors caused by the paired comparison. The significance level was set at p < 0.05.

The attachment of sensors for physiological recordings on the subject's body was time consuming for a clinical application demanding resources of the therapist and also reduced the time a subject could exercise in the Lokomat. To improve clinical applicability of the approach, it was of interest to understand, if all recorded physiological signals were necessary to perform classification of cognitive engagement or if the recorded data contained information from dependent variables. It could, for example, be possible that heart rate and breathing frequency would show a strong correlation. In this case, one of these signals could then be omitted in future recordings without degrading classification performance.

Therefore, it was investigated, which signals contained most information in terms of variance explained, and could be seen as major markers for changes in psychological states. PCA was performed for each healthy subject and each patient individually. In this analysis, five minutes data of each of the four conditions of one subject were combined before the PCA was performed. Inputs to the PCA were HR, a discrete time series of HRV, a discrete time series of the number of SCR events, SCL, skin temperature and a discrete time series of breathing frequency. PCA is a linear, orthogonal rotation, which projects the original data into a new coordinate system. In this new coordinate system, the first axis (or first principal component (PC)) explains most of the variance. The second PC explains the second most important variance, etc. The PCA is computed as

$$D^{nxm} = F^{mxm} \times A'^{nxm} \tag{3.1}$$

D is the original data, A' are the activation coefficients and F represents the loading factors. In this case, the original data consisted of a time series with n data points and $m = 6$ physical recordings (dimensions). The activation coefficients were a time series with 6 dimensions. The matrix of loading factors was the rotation matrix and defined the new coordinate system.

The number of factors k necessary to explain more than 80% of the variance was computed in all subjects ($k \in [1, n]$, where n is the dimensionality of original data, i.e. 6). A factor rotation on these first k PCs was performed to obtain a clearer picture, which input signals provided the largest variance. Factor rotation is a mathematical transformation that does not alter the subspace spanned by the

PCs, but shifts the weight of an input e.g. from the first PC to the second, while maintaining the orthogonality between the components.

Experimental Protocol

Each measurement was subdivided in a practice time with virtual environment, a baseline recording without virtual environment and the three task conditions under-challenged, challenged and over-challenged (Fig. 3.8). Although the task was designed such that it was possible for the subject to manage with little physical effort, it still involved a non-negligible physical effort. As stroke patients, a heterogeneous population, are already likely to exhibit large inter-subject variability due to impairments of cognitive and motor ability (as well as the potential effects of medications), it was decided in the experiment planning phase to avoid task order randomization in order to keep the influence of physical effort comparable. Therefore, it was decided to present the conditions in the above-mentioned order. The experiment started with the bored condition, which required the least control effort to succeed, and finished with the over-challenged condition, which required the most effort from the subjects in order to accomplish the task

The sequence of one measurement was

- Practice time: subjects became acquainted with the effects of their movements upon the system (controlling the system). The walking speed for the baseline and the balance of the measurement interval was maintained at 2 km/h. The task difficulty levels were set individually for each subject as described above.

- Five minute walking baseline: physiological signals were recorded in the Lokomat with body weight support of 30% and without the virtual environment tasks.

- Three task conditions in the virtual environment: the three task conditions were arranged in increasing levels of difficulty, each with a duration of five minutes. Five minutes was determined as a tradeoff between the time required to reach a steady state in the physiological signals and also to keep the exercise portion of the experiment time below 45 minutes. 45 minutes have been informally reported by physiotherapists to be the maximum time for patients to exercise in the Lokomat.

After the walking baseline and after each scenario, the subjects were requested to respond to the SAM. During the questionnaire response time, the virtual environment was turned off.

Subjects

17 healthy subjects (8 male and 9 female, mean age 24.1 ± 2.0 years) with no neurological and physiological impairment and 10 neurological patients (7 male

3.2 Methods

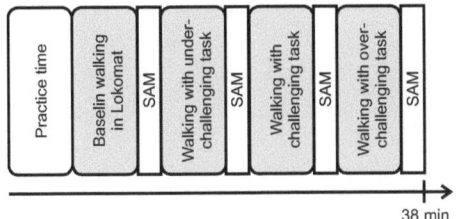

Figure 3.8: Experimental protocol

and 3 female, mean age 52.4 ± 18.9 years) participated in the study. All patients suffered from neurological gait impairment due to their pathology. Characteristics of patients that participated in the study are summarized in Tab. 3.1.

Table 3.1: Characteristics of patients recorded in experiments on the effects of pleasant and unpleasant stimuli on psychophysiological recordings. Gender: m=male, f=female. SCI= spinal cord injury.

Subj.	Sex	Age [y]	T. since inc. [m]	Lesion	FAC	WISCI II	Beta blockers
1	m	68	2.5	right middle cerebral artery stroke	4	n.a.	Yes
2	m	62	3	right middle cerebral artery stroke	1	n.a.	No
3	m	51	6.5	right middle cerebral artery and posterior cerebral artery stroke	3	n.a.	Yes
4	m	64	1	SCI ASIA A S1	n.a.	13	No
5	f	19	21	SCI ASIA C L1	n.a.	20	No
6	m	75	2.5	Myeolopathy C3-C6	n.a.	13	No
7	f	57	1.5	Guillain-Barr syndrome	0	n.a.	No
8	m	37	4.5	Hypoxic encephalopathy	1	n.a.	Yes
9	m	26	38	Subarachnoid hemorrhage Hunt and Hess grade 4-5	0	n.a.	No
10	f	65	3	right basal ganglia hemorrhage with intraventricular extension	0	n.a.	No

3.2 Methods

All patients were selected and approved for participation in the study by a clinical expert to ensure that the patients were able to follow the instructions and respond accordingly.

The study was conducted at two locations. Measurements with healthy subjects were conducted at the Spinal Cord Injury Center Balgrist, Zurich, Switzerland. Measurements with subjects who had experienced a stroke were conducted at the Neurologische Klinik Bad Aibling, Germany. Approval for both studies was obtained from local ethics committees, and all subjects or their legal representative gave written informed consent before data collection.

3.2.4 Automatic Classification of Cognitive Load

Input Data for Classification of Cognitive Load

Physiological signals were recorded as described in chapter 3.2.1 from the subject, force data from the DGO and task success data from the virtual environment. Features were extracted from the physiological data as described below, took the mean and standard deviation over 30 seconds and fused the data into one feature vector. All signal processing software was written in Matlab 2008b (The Mathworks, Natick, MA, USA, www.mathworks.com).

Force data from the DGO was weighted and summed for each step such that it reflected the current physical effort of the subjects (Banz et al., 2008). From the virtual environment, the success rate of percent correctly avoided and collected objects and percent correctly answered questions was obtained.

Virtual Task

A virtual reality task with adjustable difficulty level was used to modulate cognitive load during training sessions. The walking speed in the scenario was controlled via subject's voluntary effort in the DGO. As the DGO was position controlled, the subject could produce voluntary forces, either pushing into the movement direction of the orthosis or resisting the gait movement of the orthosis. An increase in effort yielded to an increase in virtual walking speed; a decrease in effort resulted in a decrease in virtual walking speed. While the subject could influence the virtual walking speed in the scenario, the real walking speed in the DGO was kept constant.

In the virtual task, subjects had to collect and avoid objects which were placed on a straight line and disappeared slowly in front of them. By modulation of their physical effort in the DGO, the subject could collect objects by increasing effort and avoid objects by decreasing effort. In addition to this biomechanical task, subjects had to answer questions during the task, which were displayed in a box on the screen. If the statement was correct (e.g. 1+1=2), subjects had to collect the

3 Controlling Psychological States during Robot-Assisted Gait Training

Figure 3.9: Virtual scenario. A cognitive task (question) had to be answered via biomechanical effort. If the question was correctly posed (alpha is the first letter of the Greek alphabet), then the subject had to accelerate in the virtual world and had to collect the box in front of him/her, before it disappeared. Otherwise, the subject had to decelerate in the virtual world and wait, until the box had disappeared. The time until the box disappears was coded by arrows that pointed at the next object, which made the task easier to understand. (©2010, 2011 IEEE)

box before it disappeared. If the statement was false (e.g. 1+1=3) subjects had to avoid it by decreasing the walking speed until the box disappeared (Fig. 3.9).

The task difficulty could be increased by increasing the question difficulty, by decreasing the time available to read and answer the question, by decreasing the distance between objects and increasing the time until the objects disappear. Conversely, the difficulty could be decreased by posing easier questions, allowing more time to read and react to the question, by increasing the distance between objects and decreasing the time until the objects disappeared.

Modulating Cognitive Load

Three distinct levels of cognitive load were induced by adjusting the task difficulty. In condition one, subjects were bored and under-challenged; condition two provided a cognitive challenge which was difficult, but feasible; condition three over-challenged and over-stressed subjects with an infeasible difficult task. Task difficulty was set individually: in the under-challenging condition, the task was adjusted such that subjects succeeded in over 90% of cases. The questions were very simple, the objects were placed far away and disappeared slowly such that subjects had a long time to think about the answer. In the challenging condition, question difficulty and the required reaction time were adjusted so that the success rate was between 40-80%. In the over-challenging condition, subjects had very little time to answer very difficult questions with a success rate of maximally 20%.

3.2 Methods

Correlation between Physical Effort and Cognitive Load

To verify that cognitive load and physical effort were dissociated, questionnaire results from physical effort and cognitive difficulty level were tested for significant differences between the two baseline conditions (Fig. 3.10, No task condition). It was also tested, if the change in physical effort was significantly higher than the change in cognitive load for the two baseline conditions. Both tests were performed using the Friedman test followed by a Wilcoxon test for paired comparison. To get further information, if cognitive load and perceived physical effort were dissociated, the coefficient of determination R2 was computed between cognitive load and perceived physical effort.

Setup of the Three Classifiers

A neural network, a classic linear discriminant analysis (LDA) (Bishop, 2006) classifier as well as a Kalman adaptive version of the LDA were investigated. All classifiers were trained to classify cognitive load from the recorded physiological variables and performed classification once a minute.

Neural network

For automatic classification of cognitive engagement from physiological recordings, the effectiveness of a neural network was evaluated. As a classifier, a data fitting neural network was trained (Neural Network Fitting Tool, Matlab, the Mathworks, (www.mathworks.com)), containing 30 hidden layer neurons. The neural network provided an estimation of the current state of cognitive engagement, based on the physiological recordings. 20% of the data was taken as training data, 20% as validation and 60% as testing data. As neural networks require labeled data during the training phase, the training data was labeled as 1="baseline", 2="under-challenged", 3="challenged" or 4="over-challenged". Learning was performed with the Levenberg-Marquardt back-propagation algorithm (Mor 1977). Data necessary for training of the network was selected such that an identification phase would not take longer than 20% of the whole training time in the DGO.

Classic LDA classifier

The classic LDA tries to separate two classes by mapping the input vector $\mathbf{x}$ to a one dimensional output y using the weights $\mathbf{w}$ after $y = \mathbf{xww}$ is computed by maximizing the cost function

$$J(\mathbf{w}) = \frac{\mathbf{w}'S_B\mathbf{w}}{\mathbf{w}'S_W\mathbf{w}} \qquad (3.2)$$

where S_B and S_W are the between-class covariance matrices and the within-class covariance matricies of the two classes that are to be separated. With four classes to be distinguished (no task, under-challenged, challenged, over-challenged), we trained four univariate LDA classifiers $J_i, i \in [1,2,3,4]$. The class was then identified as $max(J_i)$.

All data recorded in the 'no task' condition, regardless of the level of physical effort, was labeled as baseline to the classifier. This ensured that the classifier estimated only cognitive load and not physical effort.

Kalman adaptive LDA classifier

Kalman adaptive linear discriminant analysis (KALDA) is an adaptive version of the classic LDA classifier where the weights **w** are updated recursively using a Kalman filter when new data become available (Vidaurre et al., 2007). Every minute, when a new input vector **x** and its corresponding known output y become known, the weights **w** are updated according to the following equations:

$$H = [1, \mathbf{x}] \tag{3.3}$$
$$e = y - H \cdot \mathbf{w}_{k-1} \tag{3.4}$$
$$v = 1 - UC \tag{3.5}$$
$$Q = H \cdot A_{k-1} \cdot H^T + v \tag{3.6}$$

$$k = \frac{A_{k-1} \cdot H^T}{Q} \tag{3.7}$$
$$\mathbf{w}_k = \mathbf{w}_{k-1} + k \cdot e \tag{3.8}$$
$$\tilde{A}_k = A_{k-1} - k \cdot H \cdot A_{k-1} \tag{3.9}$$
$$A_k = \frac{trace(\tilde{A}_k) \cdot UC}{p} + \tilde{A}_k \tag{3.10}$$

where $\mathbf{w}_{k-1}$ are the old weights, **w** are the updated weights, e is the one-step prediction error, Q is the estimated prediction variance, A_{k-1} is the old a priori state error correlation matrix, A_k is the new a priori state error correlation matrix, $\tilde{A}_k$ is an intermediate value needed to compute A_k, v is the variance of the innovation process, k is the Kalman gain, UC is the update coefficient, and p is the number of elements of **w**. The starting values of A_0 and $\mathbf{w}_0$ as well as the optimal value of UC are computed from the training data set.

Originally designed for analysis of electroencephalographic data (Vidaurre et al., 2007), KALDA has already been used for two-class classification of physiological measurements and motor activity measurements in upper extremity rehabilitation (Novak et al., 2010). The original KALDA was designed for two classes, but it can be expanded to four classes by running the update process for all four classifiers

3.2 Methods

in parallel. In this case, we can either update all four classifiers for each new data point or only update the classifier corresponding to the correct class.

An important problem with the original implementation of KALDA is that it is supervised; as can be seen from equation (3), the correct output class (y) is required to update the weights. Since this information is generally not available in practice, we have modified KALDA so that it updates the weights using its own estimate of the output class rather than the correct output class. This needs to be done carefully since such an approach can also amplify errors. If an incorrect estimate is used to update the discriminant function, the discriminant function will become worse. Our method of addressing this was to generate a measure of how 'reliable' the estimate is. The system then only updates the discriminant function if the estimate is sufficiently reliable. The reliability criterion was relatively simple. As previously mentioned, the class is defined using all four classifiers as $max(J_i)$, $i \in [1,2,3,4]$. We considered the estimate to be sufficiently reliable (and updated the weights) if $max(J_i)$ was larger than all other elements of J_i by a certain reliability threshold T_R - i.e. if one of the four classifiers showed that its corresponding class was much likelier than the other three classes. The optimal value of T_R was calculated from the training data set using a sensitivity analysis.

Experimental Protocol for Classifier Training

Developing an algorithm that estimated only cognitive load required to verify that cognitive load and physical effort were dissociated. This was necessary, as the virtual task was controlled by modulation of physical effort. Data was therefore collected in which cognitive load (which was shown to be modifiable via task difficulty in chapter 3.2.3) and physical effort (required energy to walk and to control the virtual task) were not co-varied in the main protocol.

Each experimental session was initiated with a four minute baseline period, in which physical effort was varied, but no cognitive task was present ('no task' condition, Fig. 3.10). During this initial period, subjects completed two walking behaviors: passive, such that the robot provided most of the physical effort and active, over-emphasizing the gait pattern and expending additional energy.

The initial 6 minute period of only physical effort was followed by 5 minutes of exercise time, during which subjects could get acquainted with the addition of the virtual task. Meanwhile, the experimenter determined the levels of cognitive load by adjusting the distance between objects and the question difficulty level such that the task success for each condition was reached as described above.

After the baseline measurement, three different cognitive load conditions were presented in randomized order, each 2.5 minutes long (cognitive task with randomized task difficulty, Fig. 3.10). The three different levels of cognitive load were induced by adjusting the difficulty of the task at the beginning and, as necessary, during the condition such that the subjects could reach a desired task success. Dif-

3 Controlling Psychological States during Robot-Assisted Gait Training

ficulty was modulated by question difficulty, distance between objects and the time before the next object would disappear.

After baseline and after each condition, subjects were asked to answer questionnaires on cognitive load, in order to verify if subjects were really cognitively under-challenged, challenged and over-challenged. In addition, subjects were asked, how difficult they perceived the task in terms of physical effort.

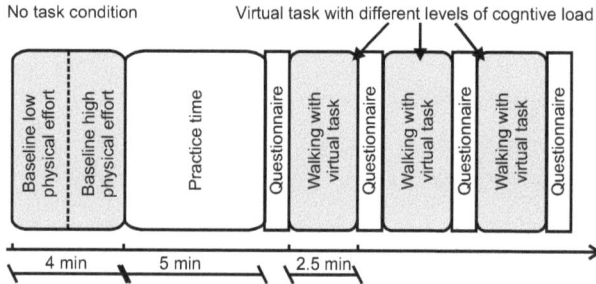

Figure 3.10: Study protocol open loop experiments. The virtual task is presented three times, each time with a different task difficulty to induce three different levels of cognitive load. The order of the conditions is randomized. (©2010, 2011 IEEE)

Performance Evaluation

The quality of the classifier training was quantified by computing percent correctly classified between the estimated and the actual cognitive load. The actual cognitive load was labeled as cognitively under-challenging, challenging but feasible and over-challenging, according to the condition the subject was in, as explained in the section "Modulating cognitive load". We investigated how well the classifier could generalize across subjects by training the classifier on all but the i-th subject and performing classification on the i-th subject, commonly called "leave one out" classification. This was done separately for the data of healthy subjects and of patients. We then cross-validated the classifier between healthy subjects and patients by training the LDA on all healthy subjects and classified the data of all patients.

To investigate, if physiological signals alone would suffice to classify cognitive load, we trained the classifier with five different input vectors:

- C1: physiological signals alone,

- C2: physiological signals with task success data from the virtual environment,

- C3: physiological signals with force data from the robot,
- C4: a joint input vector of physiological signals, task success and force data,
- C5: only task success from the virtual environment.

We compared the classification results for the five input vectors for healthy subjects and for patients. We then checked, if the KALDA would significantly improved classification compared to the classic LDA. All statistical tests were performed using the Friedman test. Afterwards, a post-hoc Wilcoxon test for paired comparison with Bonferroni correction was performed. Due to the paired comparison of five conditions, the significance level was set to 0.01.

Subjects

Open loop experiments were performed in nine healthy subjects (5 female, 4 male, 29 years ± 5) and four stroke subjects (Tab. 3.2). All subjects were naïve and had never seen the virtual environment. Subjects were excluded if cognitive impairments prevented them from reading and understanding the questions on the screen. Subjects were asked to refrain from coffee, tea and cigarette consumption four hours prior to the recording. Upon arrival, the task and the questionnaires were explained to all subjects. All subjects gave informed consent. Approval for all studies was obtained from local ethics committees, and all subjects gave written informed consent before data collection.

Table 3.2: Characteristics of patients for open loop classifier training. Gender: m=male, f=female. (©2010, 2011 IEEE)

Subj.	Sex	Age [y]	T. since inc. [m]	Lesion	Beta blockers
1	f	52	29	left ischemic infarct	no
2	m	43	5	left hemorrhagic infarct	no
3	f	37	22	left hemorrhagic infarct	no
4	m	66	29	left ischemic infarct	no

3.2.5 Closed Loop Control of Cognitive Load

Closed loop experiments with neural networks or the classic LDA were not performed, as KALDA proofed to be superior over the two other classifiers (see results).

3 Controlling Psychological States during Robot-Assisted Gait Training

Adaptation of Virtual Environment

The goal of the closed loop experiment was to reach a challenging, but feasible task difficulty for each subject, independent of the subject's abilities and the initial settings of the virtual task. We had intentionally set up a four class classifier that could distinguish between three classes of cognitive load and a baseline. The virtual environment however only allowed making the task easier or harder. We, therefore, had to reduce three classes of cognitive load to the binary decision easier or harder.

If cognitive load was classified as under-challenging, task difficulty was increased with a large adaptation step in the virtual environment. If cognitive load was classified as over-challenging, task difficulty was decreased with a large adaptation step. If cognitive load reached a state in which it was classified as challenging but feasible, the classifier evaluated if the task was by trend too easy or too hard and then also performed an update of the task difficulty, but with smaller adaptation steps (Fig. 3.11). This allowed fast convergence to a state in which the subject was cognitively challenged and prevented oscillatory behavior of the task difficulty. Theoretically, the classifier could also detect baseline. While this situation never occurred, the adaptation rules stated that no change would be undertaken in this case.

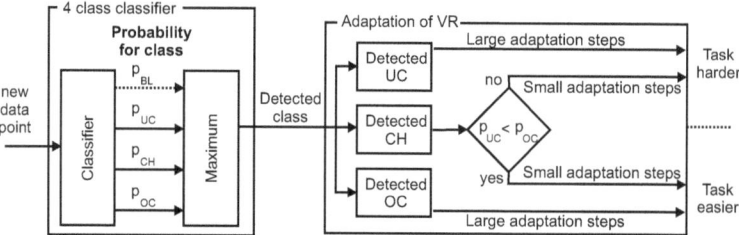

Figure 3.11: Adaptation of virtual environment based on the result of the classifier. The classifier determines the probability p for each of the four classes (BL: baseline, UC: under-challenged, CH: challenged, OC: over-challenged) and determines the current cognitive load from the largest probability. In the extreme cases (UC, OC), large adaptation steps are done when adapting the virtual task. Adaptation also happens if the classifier detects the class CH (p_{CH} larger than all other probabilities). By comparing the p_{UC} versus p_{OC}, the algorithm determines if the subject tends towards under-challenged or over-challenged and performs small adaptation steps. (©2010, 2011 IEEE)

3.2 Methods

Experimental Protocol

Subjects started to walk in the Lokomat and were given five minutes to exercise the task. The assistive force of the Lokomat was set to 100%, which corresponds to position control of the gait trajectory. Testing the controller based on physiological signals, we started the training session in an extreme condition (either too easy or too difficult), pseudo-randomized for each subject. Every 60 seconds, the classifier provided a real-time estimation of the current cognitive load, based on the last 30 seconds of data, and updated the virtual task difficulty. We allowed 10 update steps of the virtual environment, which resulted in 10 minutes Lokomat walking.

This protocol was run in two randomized conditions to evaluate the necessity of psychophysiological recordings for automatic classification of cognitive load: once, the experiment was performed with the KALDA classifier; input to the system was the full feature vector as described above. Once, only task success without additional physiological recordings was controlled to a desired level. The controller tried to set the task level difficulty such that success rate reached 70%, which was defined as challenging, but not too difficult.

Performance Evaluation of Closed Loop Experiments

While in the open loop experiments, the subjects had time to answer the questions after each condition, in the closed loop experiments we did not want to interrupt the immersion and focus on the game. We, therefore, decided against asking subjects the full set of questionnaires on cognitive load and physical effort during the closed loop experiments. Assuming that we were able to really modulate cognitive load as suggested by the results of the open loop experiments, we only tested whether or not the subjects agreed with the adaptation step of the virtual environment. The correctness of the classifier's decision was verified by asking subjects at each update step if they would want the task to be easier or harder. This meant that evaluation of the closed loop experiments was only done with two decisions (easier/harder), compared to four classes in the open loop experiments (Fig. 3.11).

While the subject's answer to the questions was not taken into account for the classifiers decision, it allowed comparison between the subjects' opinion and the adaptation steps taken in the virtual environment. Also, for comparison, the experimenter rated the performance of subjects and noted, if the task should be easier or harder from the therapeutic point of view. To avoid a bias, the experimenter rated the performance before asking the subject. Also, the experimenter could not see the classifier's decision.

Comparing the percent match between subject-classifier and experimenter-classifier was of particular interest to quantify how patients perceived and rated their own performance. While we excluded patients with cognitive impairments in this study, patients might not be able to rate their own performance subjectively compared to objective expert-rating of the therapist.

Using a Friedman test with post-hoc Wilcoxon test, we compared the closed loop system with physiological data, robot data and score information to the system that only controlled score. Significance level was set to 0.05, as no Bonferroni correction was necessary.

Subjects

Both open and closed loop experiments were performed with naïve subjects that had never seen the virtual environment before. Open loop experiments were performed in nine healthy subjects (5 female, 4 male, 29 years ± 5) and four stroke subjects (Tab. 3.2). Closed loop experiments were performed in five healthy subjects (1 female, 4 male, 32 years ± 12) and five stroke patients (Tab. 3.3). Subjects were excluded if cognitive impairments prevented them from reading and understanding the questions on the screen. Subjects were asked to refrain from coffee, tea and cigarette consumption four hours prior to the recording. Upon arrival, the task and the questionnaires were explained to all subjects. All subjects gave informed consent. Subjects were fixed into the DGO with a harness around the hip and cuffs around the legs and walked at 2km/h. For safety reasons, all subjects were connected to the body weight support system. Approval for all studies was obtained from local ethics committees, and all subjects gave written informed consent before data collection.

Table 3.3: Characteristics of patients for closed loop control of cognitive load. Gender: m=male, f=female. (©2010, 2011 IEEE)

Subj.	Sex	Age [y]	T. since inc. [m]	Lesion	Beta blockers
1	m	45	9	left ischemic infarct	no
2	m	47	14	right hemorrhagic infarct	no
3	m	63	516	right ischemic infarct	no
4	m	64	60	left ischemic infarct	no
5	f	54	54	left ischemic infarct	no

3.3 Results

3.3.1 The Effects of Pleasant and Unpleasant Stimuli on Psychophysiological Recordings

Questionnaires

None of the questionnaire answers showed a significant change between any condition. The median increased for the scenarios 'rain' and 'scary' compared to baseline. The nice scenario showed the same level of arousal as baseline. For the valence dimension only the 'rain' scenario showed a decrease compared to baseline. The 'nice' scenario and the 'scary' scenario had the same median values as baseline.

Physiological Recordings

None of the physiological recordings showed a significant change between any condition. However, the trends are reported for the sake of completeness (Tab. 3.4). Heart rate decreased compared to the baseline for the 'rain' and 'scary' scenarios. The median for the 'nice' scenarios is very similar to the baseline measurement. RMSSD showed a decrease for the 'rain' scenario compared to baseline. The values for the 'nice' scenario were very similar to the baseline measurement. LH/HF showed an increase for all visual stimuli compared to baseline. All scenarios showed higher breathing frequency compared to baseline. Skin temperature increased for all scenarios compared to baseline. The number of SCR's increased for all scenarios compared to the baseline measurement.

Table 3.4: Changes in physiological signals caused by pleasant and unpleasant stimuli. Average results of 5 healthy subjects HR=Heart rate, BF=Breathing frequency, SCR=Skin conductance responses, RMSSD=Root mean square of summed differences, ST=Skin temperature, LH/HF=low frequency components over high frequency components of heart rate frequency analysis
None of the signals showed significant differences for the different conditions

	HR [bpm]	BF [cpm]	SCR [no. of SCR/min]	RMSSD [ms]	Skin Temperature [C]	LH/HF [-]
Baseline	71.7±18.3	16.6±1.9	1.1±1.0	0.056±0.047	32.9±3.0	1.87±0.89
Nice	72.3±18.7	19.7±2.2	2.6±1.8	0.050±0.049	33.3±3.2	2.40±0.85
Rainy	69.6±18.7	19.9±3.8	2.6±1.6	0.048±0.038	33.3±2.7	2.31±1.18
Scary	69.3±17.6	18.3±3.0	2.7±2.2	0.052±0.046	33.6±1.7	2.10±0.89

3.3.2 The Effect of Virtual Environments on Psychophysiological Recordings

Questionnaires

According to the SAM questionnaires, it was possible to elicit the desired psychological state (bored, excited, over-stressed) in the majority of healthy subjects. Arousal in the four conditions baseline, under-challenged, challenged and over-challenged was significantly different among each of the conditions and increased monotonously with task difficulty levels. Similar results were found for perceived physical effort. The valence dimension demonstrated significant differences only in the over-challenged condition when compared with all other conditions. The values from the condition over-challenged were significantly lower than all other conditions (Tab. 3.5).

Table 3.5: Results of healthy subjects for the self-assessment manikin questionnaire (SAM)
Median and 95% confidence interval (CI) of the SAM dimensions (arousal, valence, and physical effort)
* Significantly different from the other conditions (baseline, under challenged, challenged, over-challenged) in this dimension ($p < 0.05$)

	Arousal	Valence	Physical effort
	Arousal	Valence	Physical effort
Baseline	1* CI: 1-2	4 CI: 4-4	1* CI: 1-1
Under-challenged	2* CI: 2-2	4 CI: 4-4	2* CI: 1-2
Challenged	4* CI: 3-4	4 CI: 3-4	3* CI: 3-3
Over-challenged	4* CI: 4-4	3* CI: 2-4	3* CI: 3-4

No significant changes between the different conditions were found in the questionnaires, neither for arousal, nor for valence nor physical effort.

Statistical Analysis of Recorded Physiological Signals

In healthy subjects, significant differences were found in several physiological signals. All results are summarized in Tab. 3.6. HR increased significantly from baseline for all conditions with the virtual task and for the conditions challenged and over-challenged compared to the condition under-challenged. The same significant changes were found for breathing frequency. Similar results were also found for the number of SCR. For all virtual task conditions, the number of SCR increased significantly compared to baseline. In addition, the number of SCR increased significantly for the over-challenged condition compared with the under-challenged condition.

3 Controlling Psychological States during Robot-Assisted Gait Training

In the time domain of HRV, RMSSD decreased significantly from baseline for the conditions challenged and over-challenged. Furthermore, a significant decrease was found for the over-challenged condition compared with the under-challenged condition. In the frequency domain no significant changes were found.

A significant decrease was also found in skin temperature. The skin temperature during conditions under-challenged and challenged was significantly decreased when compared with the baseline and the over-challenged conditions.

Compared with the very robust and variable physiological signals in healthy subjects, only three significant changes were found in patients. HR increased significantly for the challenged condition (+7.6%) and for the over-challenged condition (+6.2%) compared with baseline (median 89.7 bpm, CI: 77.5-103.2). Breathing frequency decreased significantly for the over-challenged condition (-5%) compared to the challenged condition (median 27.7 cpm, CI: 24.8-29.8).

PCA Analysis of Recorded Physiological Signals

The first two PCs explained more than 70% of the variance in all subjects, healthy as well as patients; the first three PCs explained more than 80% in all but one healthy subject and all patients (Fig. 3.12).

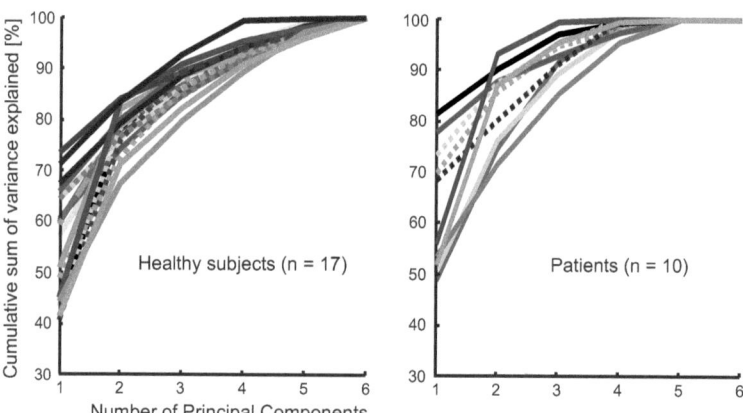

Figure 3.12: Variance explained by PCs for healthy subjects and patients. As PCA performs without loss of information, the correlation between original signal and PCA decomposed signal must be 100% if all components are combined

Via PCA, the four conditions baseline, under-challenged, challenged and over-challenged could be directly separated. The results could be visually displayed in a

3.3 Results

simple and easy to understand manner, whereas the results of the statistical analysis of the physiological data were unclear. Simplifying the display of important data reflecting the response of the patient to the exercise condition and the level of mental engagement will improve ease of use in the clinical setting and will make personalizing the exercise experience for the patient possible. A typical example is shown in Fig. 3.13 for a healthy subject and in Fig. 3.14 for a patient. Loading factor one of this particular plot relies mainly on skin temperature, while loading factor two is a combination of HR and SCL (Tab. 3.7). Each mental engagement condition may be visually distinguished from the other (as depicted by ellipsoidal boundaries in Fig. 3.13 and Fig. 3.14) providing objective information to the therapist relevant for patient therapy. The temporal evolution of the PC activation coefficients is displayed via changes in color. The lighter the color of each condition, the earlier in the condition the data was recorded. The black arrows mark the general evolution of the activation coefficients over the time course of one condition. Although classification on three PCs was performed, only two dimensions were plotted, as a 3-D plot would be difficult to display.

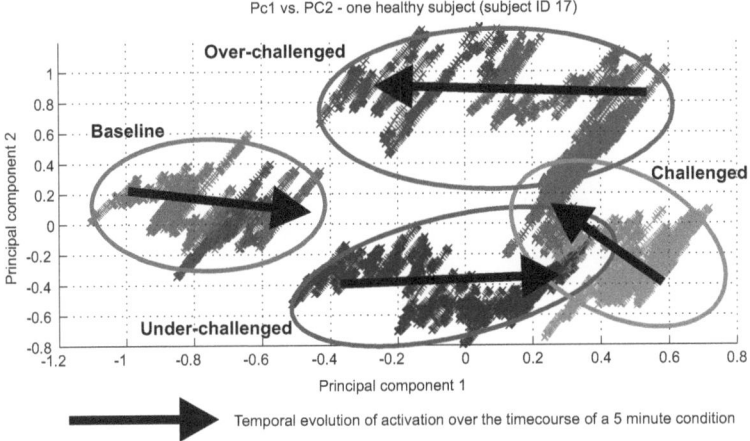

Figure 3.13: First two activation coefficients of the PCA exemplarily shown for one healthy subject (subject ID 17), separated for the conditions "baseline", "under-challenged", "challenged" and "over-challenged", plotted for the whole length of each condition (five minutes). Within one color, the darkness of the color symbolizes a later time during the condition. While the analysis was performed at once on the whole dataset, the colors illustrate the temporal evolution of the data over the four conditions.

59

3 Controlling Psychological States during Robot-Assisted Gait Training

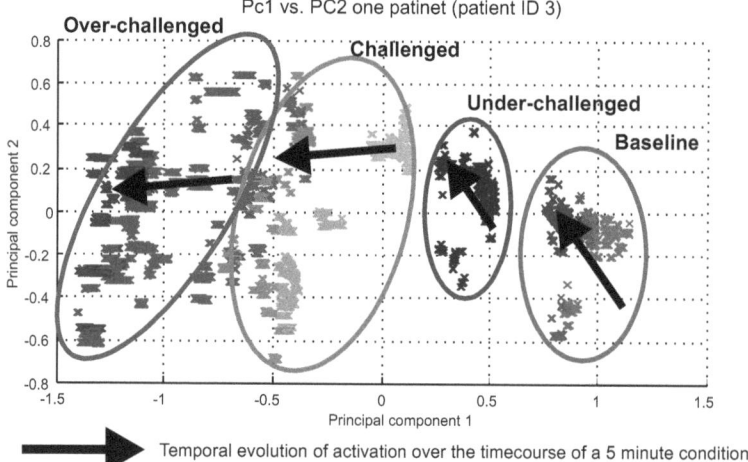

Figure 3.14: First two activation coefficients of the PCA exemplarily shown for one patient (patient ID 3), separated for the four conditions "baseline", "under-challenged", "challenged" and "over-challenged", plotted for the whole length of each condition (five minutes). Within one color, the darkness of the color symbolizes a later time during the condition. While the analysis was performed at once on the whole dataset, the colors illustrate the temporal evolution of the data over the four conditions.

Table 3.6: Statistical results of physiological recordings in healthy subjects
Median and 95% confidence interval (CI) of heart rate (HR), breathing frequency (BF), skin conductance response (SCR), square root of the mean squared differences of successive normal-to-normal intervals (RMSSD), and skin temperature
a = Significant different from the baseline ($p < 0.05$); b = Significant different from the under-challenged condition ($p < 0.05$); c = Significant different from the challenged condition ($p < 0.05$); d = Significant different from the over-challenged condition ($p < 0.05$)

	Heart Rate [bpm]	Breathing Frequency [cpm]	SCR [no. of SCR/min]	RMSSD [ms]	Skin Temperature [C]
Baseline	$73.3^{b,c,d}$ CI: 60.3-81.8	$21.6^{b,c,d}$ CI: 20.3-24.5	$0.2^{b,c,d}$ CI: 0.0-0.6	$30.0^{c,d}$ CI: 19.4-49.9	$32.5^{b,c}$ CI: 31.7-32.9
Under-challenged	$81.3^{a,c,d}$ CI: 68.0-91.2	$23.0^{a,c,d}$ CI: 22.2-26.1	$1.0^{a,d}$ CI: 0.2-3.7	27.5^{d} CI: 6.8-37.3	$30.8^{a,d}$ CI: 29.1-32.1
Challenged	$94.1^{a,b}$ CI: 77.8-103.0	$27.5^{a,b}$ CI: 24.9-30.3	3.1^{a} CI: 0.6-5.3	25.5^{a} CI: 5.0-60.2	$31.5^{a,d}$ CI: 30.5-32.5
Over-challenged	$96.4^{a,b}$ CI: 76.3-102.9	$27.8^{a,b}$ CI: 25.0-29.4	$3.3^{a,b}$ CI: 0.4-6.2	$15.3^{a,b}$ CI: 4.7-63.0	$32.0^{b,c}$ CI: 31.3-32.7

3 Controlling Psychological States during Robot-Assisted Gait Training

Table 3.7: Loading factors of the first PCs of subject 17
The physiological recordings used to extract PCs are heart rate (HR), heart rate variability (HVR), skin conductance response (SCR), skin conductance level (SCL), breathing frequency (BF) and skin temperature (ST).

	Healthy subject 17	
	PC 1	PC 2
HR	0.2471	0.6025
HRV	0.0168	0.0585
SCR	0.0002	0.0607
SCL	-0.1987	-0.7271
BF	-0.0093	-0.0668
ST	-0.9482	0.3111

Table 3.8: Loading factors of the first PCs of patient 3
The physiological recordings used to extract PCs are heart rate (HR), heart rate variability (HVR), skin conductance response (SCR), skin conductance level (SCL), breathing frequency (BF) and skin temperature (ST).

	Patient 3	
	PC 1	PC 2
HR	-0.0221	0.0284
HRV	0.0465	-0.0602
SCR	-0.3028	-0.8259
SCL	-0.6127	0.0150
BF	0.0236	-0.4651
ST	0.7278	-0.3112

Evaluation of the loading factors (i.e. the rotation matrix obtained from the PCA) of all healthy subjects and all patients revealed that PC 1 was dominated by skin temperature and SCL in both groups (Fig. 3.15). HRV played a minor role in the first three PCs for both groups. While breathing frequency was not dominant in healthy subjects, data variability of patients showed a major dependency of breathing frequency in PC 3.

3.3.3 Automatic Classification of Cognitive Load

Physical Effort and Cognitive Load

Physical effort and perceived cognitive load were dissociated in both healthy subjects and patients. While not perfectly independent, the coefficient of determination R^2 between physical effort and cognitive load was 0.22 for healthy subjects and 0.33 for patients.

In healthy subjects, at two distinct levels of physical effort without virtual task, the reported physical effort of healthy subjects increased significantly for the harder physical effort condition compared to the easier physical effort condition ($p < 0.003$). Meanwhile the reported cognitive load increased, but not significantly (Fig. 3.16, top left) from easy to hard physical effort. While perceived physical effort and perceived cognitive load both increased, the increase in physical effort was significantly higher compared to the increase in cognitive load ($p < 0.02$). Conversely, in the cognitive task conditions, the reported physical effort did not increase significantly while the perceived cognitive load increased significantly from the under-challenged to the over-challenged condition ($p < 0.02$, Fig. 3.16, top right). Again, perceived physical effort and perceived cognitive load both increased.

3 Controlling Psychological States during Robot-Assisted Gait Training

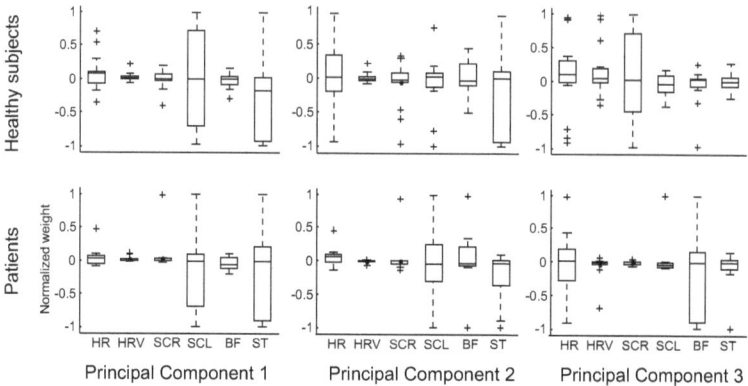

Figure 3.15: Comparison between the loading factors of the first three PCs of healthy (n=17) subjects and patients (n=10). The physiological recordings used to extract PCs are heart rate (HR), heart rate variability (HVR), skin conductance response (SCR), skin conductance level (SCL), breathing frequency (BF) and skin temperature (ST).

However, in the cognitive task condition, the increase in cognitive load was significantly higher compared to the increase in physical effort ($p < 0.005$).

In patients, changes in perceived physical effort and perceived cognitive load showed similar trends as in healthy subjects. None of the results in patients were statistically significant, which could possibly be attributed to the small sample size. The change in physical effort during the no task condition increased, but not significantly ($p < 0.13$, Fig. 3.16, bottom left). The increase in physical effort was almost significantly higher compared to the increase in cognitive load ($p < 0.08$). In the condition with cognitive task, cognitive load increased, but again not significantly ($p < 0.2$, Fig. 3.16 bottom right).

Classification Performance of Training Experiments

The classification results from all classification experiments for healthy subjects and patients are summarized in Tab. 3.9. With four classes, 25% correct classification would correspond to chance. In healthy subject, the classic LDA as well as the KALDA only reached a level slightly above chance when used on physiological signals alone. Physiological signals in combination with force data from the robot did not improve classification results of the classic LDA; the Kalman filter however could improve the classification based on physiological signals and force data by 12%. A key input was the task success in the virtual environment (score), which raised the classification results to 88% correct classification. In patients, the LDA performed equally poor in data sets with physiological signals alone or physiology signals in combination with robot force data. However, the patients could benefit to a much larger extend from the KALDA approach compared to healthy subjects, as improvements of up to 25% correct classification could be achieved. Interestingly, in patients, score information from the virtual environment could not improve classification results of the KALDA classifier.

Table 3.9: Open loop, "leave one out" classification results for healthy subjects and patients for four different kinds of input vectors. Results are presented as percent correctly classified.
Phys.: Physiological signals, score: score information from the virtual environment, force: force signals from the robot. (©2010, 2011 IEEE)

		Input vector to classifier				
		All	Phys. and score	Phys. and force	Phys. only	Score only
Healthy subjects	Neural Networks	85±8	86±11	50±15	36±15	77±7
	Classic LDA	88 ± 9	84 ±10	46±17	34±18	74 ± 7
	KALDA	88±11	85±10	58±11	38±18	75±6
Patients	Neural Networks	70±11	65±7	58±14	51±20	59±6
	Classic LDA	60±16	65±25	50±38	45±41	57±5
	KALDA	75±26	70±26	75±10	70±12	60±4

In healthy subjects and patients, the statistical tests showed a significant improvement of classification results if score information was present. Classifier C1 and C2 had score information available and classified cognitive load better than classifiers C3 and C4 that did not have score information available (p ¡ 0.007). The KALDA did not significantly improve the classification results neither in healthy subjects (p=0.89), nor in patients (p=0.11). The classifier with all information available (C1) did classify cognitive load better than the classifier base only on score (C5). However, results were not statistically significant after the Bonferroni correction was applied (p = 0.026 in healthy subject, p = 0.014 in patients).

Although only two significant differences were found in all physiological recordings over all conditions of patient data, the classification of the different psychological states using a neural network was possible for healthy subjects and also for subjects who had suffered from stroke. As described in the methods section, classification results or a neural network was evaluated for two different sets of input data: on the one side with six physiological parameters extracted, on the other side using only the physiological signals dominant in the first three PCs. Mean classification error was 1.4% for the full and 2.5% for the reduced dataset in healthy subjects and 2.1% for the full and 4.7% for the reduced dataset for patients (Tab. 3.9).

3.3.4 Classifier Performance during Closed Loop Control of Cognitive Load

On average, classification of cognitive load in healthy subjects during closed loop control was achieved with 87% ± 8% correct classification. As explained above, this number refers to the percent match of the classifiers result with the questionnaire answer of the subject. In healthy subjects, the experimenter rating of cognitive load coincided to over 95% with the decision of the subject. This system used a joint input vector of physiological data, force data from the robot and task success data from the virtual environment in combination with the KALDA classifier. Controlling only task success without the use of physiological signals, an average classification result of 64% ± 11% correctly classified could be obtained (Tab. 3.10).

In patients, the percent match between the patients decision and the KALDA classifier was very low However, the experimenter matched the decision of the classifier with 80% (Tab. 3.10). In healthy subjects, the closed loop system with physiological data, robot data and score data performed significantly better than the system based only on score information ($p = 0.039$).

Table 3.10: Results of closed loop experiments in healthy subjects and patients. Results are presented as percent match between the decision of the classifier and the decision of subject or experimenter. Although the subjects and experimenters were asked for their personal rating, this information did not influence the adaptation of the closed loop system. (©2010, 2011 IEEE)

		Healthy subjects Mean ± std	Patients Mean ± std
With physiological data	Subject-classifier	87±8	53±33
	Experimenter-classifier	85±10	80±8
Only task success (score)	Subject-classifier	64±11	68±17
	Experimenter-classifier	68±12	74±17

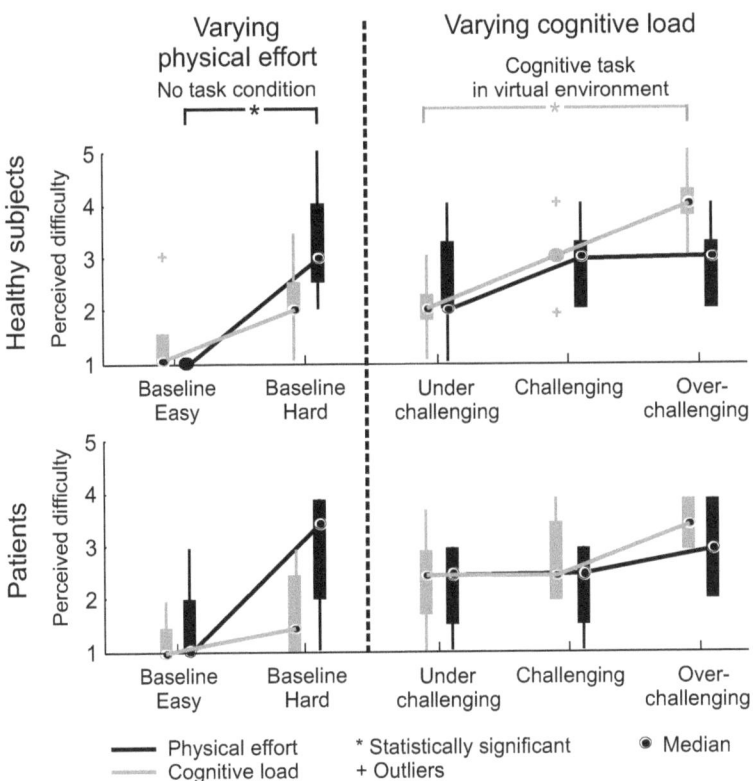

Figure 3.16: Boxplots of perceived physical effort and cognitive load in healthy subjects and patients. The graph shows median and 25%-75% percentiles of questionnaire answers. (©2010, 2011 IEEE)

3.4 Discussion

VR-enhanced robot assisted treadmill training was performed in healthy subjects and neurological patients. It was possible to demonstrate that psychophysiological measurements alone were sufficient to detect the current psychological state of a subject. The difficulty of a virtual task during the rehabilitation training was parametrically manipulated; three levels of difficulty were established including under-challenging, challenging and over-challenging to solicit an experience of being bored, of being excited, or of over-stress (as displayed in Fig. 3.1). ECG, breathing frequency, skin temperature and the galvanic skin response were recorded. HR, SCL and skin temperature were found to be usable as markers for psychological states in the presence of physical effort (corresponding very well with the gold standard instrument to measure psychological state, the SAM), associated with the exercise typically used for patients who have suffered from neurological injury.

Bio-cooperative closed loop control of cognitive load was then performed during robot assisted gait training in healthy subjects and neurological patients after stroke. Using psychophysiological measurements, robot force data and performance data from a virtual environment from open loop experiments, three different classifiers were then trained: a neural network, a linear classifier and an adaptive linear classifier. As stroke is reported to cause disturbances in autonomic functions and therefore in physiological signals (Korpelainen et al., 1996; Tokgozoglu et al., 1999; Meyer et al., 2004), the KALDA was used, which automatically adapted the classifier to the physiological responses of subjects. The KALDA generated data labels based upon its own class probabilities and therefore did not need any further input from therapist or subject.

3.4.1 Pleasant and Unpleasant Stimuli in Psychophysiological Recordings

The pleasant, unpleasant and scary virtual scenarios had an influence on subjects' physiological signals and self-reported level of arousal and valence. However, the changes in physiological signals did not show a clear picture. A clear statement of these results is difficult as the number of subjects was small and the variability in physiological response was higher as expected. This could be an explanation why a distinction between the different scenarios, via the measured physiological changes, is not possible. However, the effects were negligible compared to the effects of the cognitive tasks evaluated later on.

3.4.2 The Effect of Virtual Environments on Psychophysiological Recordings

Questionnaires

Evaluation of questionnaires from healthy subjects confirmed that virtual tasks of different difficulty levels can indeed result in different levels of mental engagement, i.e. of being bored, challenged or over-stressed. Also in healthy subjects, descriptive statistics would suffice to distinguish between different levels of mental engagement: the automatic classification worked in all but two subjects with less than 2% classification error.

In patients, neither questionnaires nor physiological signals showed a picture as clearly as in healthy subjects. While therapists anecdotally reported that the virtual task bored, challenged or over-stressed the patients, the questionnaires did not confirm this observation. The Lokomat device is typically used in gait rehabilitation of patients with little to no walking ability. In order to ensure that the approach is clinically applicable in the future, data was explicitly from severely affected patients. A possible explanation of the poor results from questionnaires is that patients with cerebral lesions might suffer from cognitive deficits, which might prevent them from assessing, expressing, and verbalizing their level of mental engagement during rehabilitation. This was consistent with reports by the therapists. In addition, it was reported that patients did not usually admit if they experienced the task as too difficult, perhaps because they were very ambitious and also determined to successfully solve the task. Although the questionnaire showed only few changes, the predefined scores (success rate of 100% for the under-challenged condition, of 80-90% for the challenged condition and of 10-20% for the over-challenged condition) were achieved by every patient. Another explanation might be the fact that walking with the help of the DGO was an experience that is positively perceived by patients who may be otherwise unable to walk well on their own.

Principle Component Analysis

We applied a PCA to identify the minimum set of physiological signals that would be necessary to perform classification of mental engagement while not degrading the performance of the classifier. In both, healthy subjects and patients, skin temperature and SCL were the main psycho-psychological responders to the intervention. Also, HRV did not contribute significantly to the first three PCs. As discussed above, HRV might have been reduced due to the physical effort involved in walking.

Relevant for clinical use are the results in patients: the loading factor of HR only occurred with a noticeable magnitude in the third PC. As HRV and HR, which are both computed from the ECG, did not significantly contribute to the variance of the recorded data in patients, the ECG might not be required in future experiments exploring mental engagement. Classification results of a dataset containing only the data dominant in the first three PCs confirmed that classification performance dropped by less than one percent. While skin temperature and SCL sensors are easily attached to the patients' unaffected finger, attachment of the ECG electrodes requires more time. In clinical day to day use of the system, reducing the complexity of data collection in this manner would shorten the time necessary to prepare a patient for therapy.

Study Limitations

There are three limitations of this study: lack of task order randomization, the heterogeneous group of patients, and differences between the VR setups of patients and healthy subjects. First, the order in which the control and the patient group performed the tasks was not randomized. Hence the results of the physiological recordings may have been influenced by the task order. From a methodological viewpoint, a fully randomized task order would have been preferable but I consciously decided to use a non-randomized order. This ensured maintaining a comparable level of physical effort among subjects. In addition, subjects with neurological gait impairments are assumed to already show a large inter-subject variability due to their motor- and cognitive impairments. Although a task was chosen in which higher task level difficulty could be achieved with little increase in physical effort, a task order randomization would have certainly further increased the variability in performance measures among subjects. Thus, it can be suspected that a further increased variability could have made a comparison between subjects impossible and might hinder uncovering potentially important results.

Second, the patient group was composed of patients with different neurological gait impairments, some of them taking medication that altered physiological responses such as beta blocking medicine. In order to get statistically significant physiological data, a homogeneous group of stroke patient who were not taking beta blocking medication would have been preferable. I decided to establish mental state classification in a broad variety of patient groups reflecting the heterogeneous population of realistic clinical patients since the goal of this work was designing and developing a practical tool for use in a clinical setting., Therefore, a variety of subjects were included in terms of lesion (stroke, SCI) and medication (beta blocking medication), in order to cover the whole range of possible patient characteristics that might be present in a rehabilitation environment, and who might benefit from a combination of a DGO with virtual environments.

Third, a larger screen had to be used for display of the virtual environment for the healthy subjects than for the patients, as the back-projection screen was installed in front of the Lokomat at Balgrist University Hospital, where recordings with healthy subjects were conducted. The larger screen size potentially elicited stronger responses in healthy subjects compared to neurological patients, which

3.4 Discussion

saw the virtual environment on a 42 inch screen. This could explain the fewer significantly different results in physiological recordings of patients compared to healthy subjects due to a lower experience of immersion.

3.4.3 Classification Performance

In open loop experiments, 88% correct classification was possible in healthy subjects. The KALDA could only further improve the results of classification that did not rely on the score from the virtual environment. However, in patients, the KALDA allowed for up to 75% correct classification. Possibly, a broader basis of patient data would have even further improved the results. The score information from the virtual environment was a key input for the classifier, in healthy subjects and patients alike. This appears logical, as the different levels of cognitive load were induced by adjusting task difficulty via task success. The neural network needed an initial training phase, after which Matlab provided a Simulink model of the network which was real time capable. The identification phase was expected to last only few minutes, which would be reasonable in a clinical setting when compared to a normal Lokomat exercise session time of 45 minutes. However, in the light of the KALDA results, neural networks appear to be unsuitable for real time classification and control of cognitive load.

While closed loop control of cognitive load could be achieved with 88% correct predictions in healthy subjects, patient results show only 53% correct classification (Tab. 3.10). When comparing the results of the classifier with the information obtained from asking the patients, the controller only performed 3% above chance level, as subjects had been asked only if they wanted the task to be easier or harder. However, taking into account possibly decreased self assessment capabilities, the reported answers of patients did often not reflect the objective assessment of the therapist.

In this light, it can be argued that the 80% ± 8% of correct match between the classifiers decision compared to the experimenters rating reflects the capabilities of the classifier more realistically (Tab. 3.10). Patient 4 for example started the experiment with a virtual task that over-challenged him. Although he obtained a score of 0% in the first three minutes, he wanted the task to be more difficult.

3.4.4 Necessity for Kalman Filters in Neurological Patients

A disturbance of the autonomic functions was often described to affect physiological processes in cerebro-vascular diseases (Korpelainen et al., 1996; Tokgozoglu et al., 1999; Meyer et al., 2004). In this context, a decrease in HRV (standard deviation of RR intervals, low frequency and high frequency) was found for stroke patients (Korpelainen et al., 1996; Tokgozoglu et al., 1999). In addition, it was shown that skin temperature was lower on the contralesional side after stroke (Naver et al.,

1995) and that the sympathetic skin conductance was altered in amplitude and delay (Korpelainen et al., 1993). Furthermore, medication of stroke patients can influence physiological signals, as for example beta blockers, which alter the cardiovascular response to psychological or physical stress.

The classical LDA only reached open loop classification result of 45%-65% correct classification (Tab. 3.9) due to the large variety of possible changes in physiological responses compared to healthy subjects. Also, the task might have been physically much more demanding for patients, which would alter effort related physiological signals such as mean heart rate, heart rate variability or breathing frequency. The Kalman adaptive classifier (KALDA) could take alterations of physiological signals caused by the stroke into account and improved the classification by up to 25% to a maximal classification result of 75%. Note that, with four classes, 25% classification correctness would correspond to chance.

The efficacy of the KALDA for classification of neurological patients became even more apparent when compared to classification results of healthy subjects: the classifier, trained on healthy and patient data, could achieve 88% correct classifier even in its none-adaptive version. Possibly, this is due to the fact that most healthy subjects responded to the intervention in a similar manner.

3.4.5 Transfer to Daily Clinical Routine

From the viewpoint of clinical applicability, a classifier based on task performance alone would be preferable compared to a classifier which also included physiological signals. Measuring physiological signals always included sensor placement, which was time consuming for the experimenter and uncomfortable for the subject. In particular, the time to place the sensors will be an issue during rehabilitation, as training-time is limited. With a classifier based on task performance, no further physiological measurements such as ECG, skin conductance, breathing or skin temperature would be necessary. All information needed for the classification would be provided by the virtual environment.

The necessity of physiological signals was therefore examined for classification of cognitive load compared to a controller that adapted the virtual environment solely based on the subject's performance in the task (Tab. 3.10). The score alone provided 74% correct classification in healthy subjects and 60% correct classification in patients. The physiological signals therefore only improved the results by 13% and 15% in healthy subjects and patients, respectively (Tab. 3.9). While this is only a small improvement compared to the additional effort of attaching the sensors, the importance of physiological data in the decision process of classification is exemplified with data from healthy subject 4 (Fig. 3.17). The classifier only takes the last 30 seconds of data into account (dashed box in plot). The classifier (C) with physiological signals as input decided in decision 1 to make the task easier, and in decision 3 to make the task harder. In both cases, the classifier's decision

3.4 Discussion

coincided with the decision of the subject (S). The score controller would have decided in both cases to make the task harder, as the optimal range of task success (>70%) was not yet reached and the data for the decision, acquired in the last 30 seconds, was similar in both cases. Therefore, the physiological data provided the deciding information on the cognitive load of the subject.

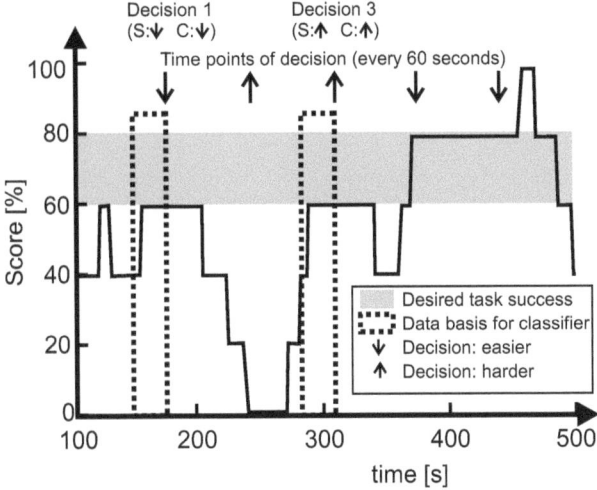

Figure 3.17: Exemplary plot from data of healthy subject 4 from closed loop control of cognitive load using the KALDA classifier with physiological input. C: classifiers decision, S: subjects decision. (©2010, 2011 IEEE)

This shows that for control of cognitive load, physiological signals are a necessary source of information required by the classifier. The score-based classifier showed good performance for extreme conditions under-challenged and over-challenged, but its performance dropped for situations, in which the subject reached a challenged state. However, for clinical use, down-scaling of the required physiological signals might provide a good tradeoff between effort for therapeutic staff involved with attachment of sensors and the benefit of assessing cognitive load.

However, given the drawbacks of the additional time required for sensory placement and the additional financial investment in purchasing the sensory equipment, a seamless integration of the presented approaches on cognitive control into daily clinical practice seems unlikely in the near future. Particularly breathing and ECG sensors are difficult to attach to patients. The quality of ECG recordings in particular can suffer greatly from artifacts induced by friction between harness and electrode. The breathing signal gets disturbed if the patient is talking or coughing. Careful monitoring of ECG and breathing signal are therefore required. This however reduces the attention the therapist can provide for the patient.

New methods for adaptation of task difficulty based on task performance are currently evaluated for automatic task adaptation. These methods are based on Fitt's law (Fitts, 1954), and have been used for automatic adaptation of task difficulty (MacKenzie, 1995) in healthy subjects. In short, Fitt's law proposes an almost linear relationship between the difficulty of a task (easy to hard) and the time required by the subject for task execution (short to long), where the perception of task difficulty and the execution time are subject individual and need calibration in the beginning. As execution time is known from the robot data and the virtual environment, task difficulty can be estimated based on an initial calibration. These methods are currently evaluated for task difficulty adaptation in stroke rehabilitation (unpublished work, Lukas Zimmerli, Hocoma Ag, Volketswil, Switzerland).

3.4.6 The Influence of Physical Activity, Neurological Deficits and Medication

To my best knowledge, no one has previously performed estimation of psychophysiological states in neurological patients during walking. This work explicitly takes into account the effect of physical activity caused by walking as well as changes in autonomous responses of the CNS in patients.

While HRV was previously used as a major marker for stress (Rani et al., 2002), HRV in our results did not change significantly with changing levels of stress. Although the physical effort, necessary to control the virtual task, was kept as low as possible, the statistical evaluation confirmed that HRV did not vary significantly. As HRV has been also shown to decrease with increased physical effort such as treadmill walking (Bernardi et al., 1996), Lokomat walking might have occluded psychological influence of stress on HRV.

The lack of statistically significant changes in physiological responses of patients might be explained by the effects of the lesion. In stroke patients, a disturbance in the autonomic functions caused by the lesion was often described to affect physiological processes in cerebro-vascular diseases (Korpelainen et al., 1996; Tokgozoglu et al., 1999; Meyer et al., 2004). In this context a decrease in HRV, i.e., standard deviation of RR intervals, low frequency and high frequency was found for stroke patients (Korpelainen et al., 1996; Tokgozoglu et al., 1999). In addition, it was shown that skin temperature is lower on the contralesional side after stroke (Naver et al., 1995) and that the sympathetic skin responses were altered in amplitude and delay (Korpelainen et al., 1996). Furthermore, medication of stroke patients can influence physiological signals, as for example beta blockers, which influence

3.4 Discussion

the cardio-vascular response. In spinal cord injury patients with cervical lesions, autonomic nervous control might be disturbed as well, as the sympathetic nerve fibers that innervate the heart leave the CNS between T1 and T4 with implications for the design of applications of these measures in real time for such patient populations.

Medication might have altered autonomic nervous control in addition to changes caused by the lesion. Three of ten patients (Tab. 3.1) received beta blocking medication at the time of recording, which are known to limit heart rate adaptation to increased physical stress. The average classification error (Tab. 3.9) might have been lower, if patients on beta blocking medication had been excluded. Nevertheless, such patients were included, as exclusion would have weakened the usability of the system in a clinical setting outside a lab environment.

3.4.7 Dissociation of Physical Effort and Cognitive Load

We could show that physical effort and cognitive load were dissociated. If they had not been dissociated, the classifier could have classified physical effort instead of cognitive load. While the dissociation of perceived physical effort and perceived cognitive load was not perfect, the change in physical effort was significantly larger than the change in cognitive load during the 'no task' condition. Conversely, the change in cognitive load from under-challenged to over-challenged was significantly larger than the change in physical effort (Fig. 3.16).

This verified that the classifier did indeed classify cognitive load, despite the similar trends of both quantities. Heart rate, HRV, breathing frequency and skin temperature, are influenced by physical effort. This was taken into account in the choice of the baseline measurement. During the baseline measurement, subjects had to vary their physical effort (Fig. 3.10, 'no task condition'). This data was labeled as "baseline" to the classifier. Training the classifier on data that did not include the initial baseline measurement, the classification performance dropped to values below 30% (results not reported).

3.4.8 Extension of the Approach to Upper Limbs

Our approach is not limited to a particular gait orthosis, and is also not limited to rehabilitation of the lower limbs. In robot assisted arm rehabilitation, as performed with the ARMin (Nef et al., 2007, 2009), the HapticMaster (Houtsma and Van Houten, 2006) or the MIT Manus (Aisen et al., 1997), the lower level of physical effort (as compared to walking) might even improve the results obtained from healthy as well as neurologically impaired subjects.

3.5 Contribution

The methods presented in this chapter were developed as part of my thesis. The open loop patient data of the effect of virtual environments was recorded by my colleagues Dr. Mark Sapa, Jeannine Bergmann and Carmen Krewer from the 'Neurologische Klinik Bad Aibling'. The final versions of the virtual environment was programmed by Lukas Zimmerli. The Kalman Adaptive Linear Discriminant Analysis was first implemented by Domen Novak under supervision of Prof. Dr. Matjaz Mihelj and Prof. Dr. Marko Munih, all from University of Ljubljana. The offline extraction algorithm for the GSR signal was a contribution of Prof. Dr. Mel Slater from University of Barcelona. An offline version of a HR extraction algorithm from ECG was contributed by Dr. Johannes Schumm from ETH Zurich. Prof. Dr. Eric Perreault advised me on the dissociation between physical effort and cognitive load.

I recorded all data from healthy subjects as well as the closed loop data in patients and performed the analysis of all data. I investigated PCA and neural networks as potential classifiers, implemented the open loop LDA and participated in the implementation of the KALDA classifier. I programmed all online versions of the feature extraction algorithms and computed all statistics. I programmed the adaptation systems for the virtual task.

4 Monitoring Heart Rate During Control of Physical Effort

4.1 Introduction

Monitoring heart rate during physical exercise can be crucial to prevent over training, which would decrease the efficacy of the rehabilitation training (Achten and Jeukendrup, 2003) and might become dangerous to the patient. We, therefore, developed a model of heart rate as a function of robot-assisted gait training which could be used to predict the temporal evolution of mean heart rate and prevent dangerous situations before they occurred. We first analyzed the settings of a gait robot, which would influence heart rate. As power expenditure of a subject was shown to have major influence on heart rate, we analyzed where power could be exchanged between an exoskeleton gait robot and a subject. To parameterize a model, we then performed one experiment with five healthy subjects and two experiments with a total of eight stroke patients. The model identification in healthy subjects and patients have been reported previously in a conference paper (Koenig et al., 2009). The text used from (Koenig et al., 2009) is ©2009 IEEE, reprinted with permission.

4.2 Methods

4.2.1 Power Exchange during Lokomat Walking

During robot assisted gait training, the robot can exert large forces onto the patient's legs to guide them on a reference trajectory. This power exchange between the device and the patient has a major effect on the heart rate. At high guidance forces, i.e. with a stiff impedance controller, the patients have the possibility to walk actively, i.e. pushing into the orthosis with high forces, or behave passively, letting the robot move their legs.

The power expenditure of a subject during exercise on a bicycle ergometer (Vokac et al., 1975; Borg et al., 1987; Arts and Kuipers, 1994) and during arm cranking (Borg et al., 1987) was reported to correlate approximately linearly with heart rate. Increased power expenditure during robot-assisted walking was observed to result in increased heart rate. In order to evaluate the ports where power expenditure can take place, power exchange in terms of forces and torques between human and robot were analyzed. This can happen with three different components of the robotic system: with the actuated orthosis, with the body weight support system, and with the treadmill/ground. (Fig. 4.1). Not all forces are measurable in real-time, but not all of them lead to power exchange either.

As the Lokomat gait orthosis was used for our experiments, this analysis is tailored to the Lokomat system. Still, this analysis is applicable for any exoskeleton gait robot equipped with a body weight support system that helps patients walking on a treadmill or overground such as the WalkTrainer (Stauffer et al., 2009), the LOPES (Veneman et al., 2007) or the Autoambulator (www.autoambulator.com).

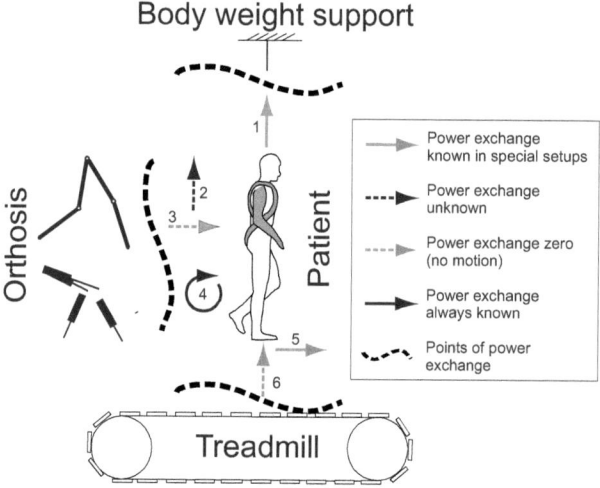

Figure 4.1: Forces/torques and corresponding power exchange between Lokomat and patient. (©2009 IEEE)

Most importantly, the Lokomat delivers joint torques to the human, which are generated by the orthosis' drives (arrow 4, Fig. 4.1). These torques are responsible for the main amount of power exchange. The vertical displacement and, therefore, the power exchanged with the body-weight support system, is small and can therefore be neglected. If the system is equipped with an active body weight support system (the commercially available Lokolift), this active support system could be used to measure the applied vertical force, and the power could be calculated with an additional sensor for vertical motion (arrow 1, Fig. 4.1). In the experimental

4.2 Methods

setup, a passive body weight support system was used. The forces between human and treadmill can be separated into horizontal (arrow 5, Fig. 4.1) and vertical (arrow 6, Fig. 4.1) ground reaction forces. The vertical ground reaction forces can be measured when the treadmill is equipped with force sensors, but do not cause power exchange as there is no vertical displacement of the contact point. The horizontal interaction forces, i.e. the shear forces, allow exchange of power between human and treadmill. To quantify the magnitude, pre-tests were conducted in which subjects walk passively while the synchronization of Lokomat orthosis and treadmill was altered to create large shear forces. The tests showed that the forces between human and treadmill increased the amount of power exchanged with the orthosis by a maximum of 10%; the exchanged power with the treadmill was therefore neglected in the following. A treadmill equipped with shear force sensors would allow for a more exact calculation. Finally, there are also vertical and horizontal forces exchanged with the orthosis and the backrest (arrow 2 and 3, Fig. 4.1). Horizontal forces do not lead to a power exchange, as the position of the human is stationary to the backrest in the horizontal direction. Vertical forces would be difficult to quantify, but as vertical displacement is small, the power exchanged here is neglected.

In summary, torques exchanged between human and orthosis (arrow 4, Fig. 4.1) were considered as the dominant port for power exchange in the system of Fig. 1.1. Due to the sensor placement in the Lokomat, only torques exchanged between the Lokomat's drives and the exoskeleton could be recorded. Using the recorded gait trajectory, the torques necessary to move the exoskeleton on this trajectory can be computed and subtracted from the recorded torques. This permits to determine the torques and thereby the power exchanged between Lokomat and human.

The power in the Lokomat during walking ($P_{Lokomat}$) can be computed as

$$P_{Lokomat} = \tau^T \dot{q}, \tag{4.1}$$

where $\tau = [\tau_{hip\ left}\ \tau_{knee\ left}\ \tau_{hip\ right}\ \tau_{knee\ right}]^T$ is the interaction torque between the human and the Lokomat and $\dot{q} = [\dot{q}_{hip\ left}\ \dot{q}_{knee\ left}\ \dot{q}_{hip\ right}\ \dot{q}_{knee\ right}]^T$ is the angular velocity of the orthosis. This power gives an indication how active the human is. Positive values indicate that the human walks actively. Negative power means that the Lokomat assists the human.

4.2.2 Model Identification in Healthy Subjects

The electrocardiogram (ECG) was recorded in five healthy individuals to define a model for the cardiovascular process of subjects during Lokomat walking (3 m and 2 f, 25.0 yr ± 2.3, 77.2 kg ± 8.0). The ECG was recorded with a gTec (www.gtec.at) amplifier, sampled at 512 Hz, filtered with a 50 Hz notch filter and bandpassed with a 20-50 Hz Butterworth filter of 4th order. HR was then extracted in real time using a custom steep slope detection algorithm. All software was implemented in Matlab

2008b (www.mathworks.com). The study protocol was approved by local ethics committees and subjects gave informed consent.

Three Lokomat parameters were varied: treadmill speed, GF and BWS. For experiments in healthy subjects, a velocity-dependent friction force was implemented to investigate the effects of changes in power expenditure. Friction was computed as

$$F_{friction} = \alpha v_{TM}^2 \qquad (4.2)$$

where v_{TM}^2 is the treadmill speed and α is a scaling factor. The friction caused additional power expenditure adding up to the expenditure related to walking. Subjects walked at three walking speeds [1, 2 and 3 km/h], three GFs [0%, 50% and 100%], three different levels of BWS [0%, 30% and 60%] and three different levels of friction ($\alpha = 0$, 0.5 and 1). Note that a GF of 100 % meant a maximally stiff impedance controller and 0 % a fully transparent orthosis. Maximal gait speed of 3 km/h was chosen, as the Lokomat is limited to 3.2 km/h for patient safety.

The dependency between walking and heart rate of healthy subjects has been previously investigated. Increases in treadmill speed were shown to linearly increase heart rate (Fagraeus and Linnarsson, 1976; Hajek et al., 1980; Pierpont et al., 2000; Achten and Jeukendrup, 2003). This can be interpreted as lowpass-like reaction to a sudden increase of oxygen demand, which we modeled as a first order delay (PT) element ($\frac{k}{\tau s+1}$). Treadmill acceleration resulted in an overshoot in heart rate before steady state was reached (Fagraeus and Linnarsson, 1976; Hajek et al., 1980; Baum et al., 1992). An undershoot was observed after a negative acceleration. Holmgreen reported a drop in arterial pressure that reached its minimum 10 seconds after onset of exercise (Holmgren, 1956). The heart rate overshoot might be caused by a first overreaction of the cardiovascular system to compensate for the blood pressure drop. Feroldi et al. argued that the overshoot might be a result of changes in the balance between sympathetic and parasympathetic activity (Feroldi et al., 1992). The overshoot and undershoot behavior was modeled as a second order derivative (DT) element ($\frac{ks}{(s+\tau)^2}$, Fig. 4.2). The power expenditure of the human was taken as a linear input parameter modeled as a first order PT element. After longer training durations a fatigue effect, which resulted in increased heart rate at steady state, was observed and described by several researchers (Fagraeus and Linnarsson, 1976; Hajek et al., 1980; Su et al., 2007b). This was modeled as a first order lowpass element. This resulted in a model with five scaling factors and six parameters (Fig. 4.2).

Estimation of the the scaling factors and parameters for each subject were done using a genetic algorithm in combination with a gradient descent optimization, as we wanted to explore the whole parameter space for solutions. Validation of the model was performed with a velocity profile different from the one used for model identification (Fig. 4.4). The goodness of fit was assessed with the coefficient of determination, R^2.

4.2 Methods

For real-time heart rate prediction, identification of the subject specific parameters needed to be repeated at each training. Online identification of subject specific parameters was performed, optimizing only over the first 12 minutes of the speed profile in (Fig. 4.4). After the first 12 minutes, the model parameters were fixed and used for heart rate prediction during training sessions of 37 minutes.

Estimating power exchange during Lokomat walking as described above could not be performed with 100% accuracy, as no sensor measured the shear forces between feet and treadmill. In pre-tests, the synchronization was altered between treadmill and Lokomat, thereby generating large shear forces between the subject's foot and the treadmill. Although we could not directly measure these shear forces due to the inability to detect them, we computed power for both cases (large and small shear forces) and found a maximal deviation of ±10% of power for both cases. To ensure the validity of our model even for extreme cases of vertical ground reaction forces, we first optimized model parameters over the recorded data as described above, then set the parameter values fixed and reduced / increased the power signal by ±10%. We then recomputed R^2 values for the heart rate prediction and compared how much the prediction quality degraded from possible inaccuracy of sensory data.

4.2.3 Model Identification in Patients

Parameter estimation was performed with eight stroke patients (Tab. 4.1) by looking at steady state increases in heart rate from baseline heart rate measured during standing. Investigating guidance force, treadmill speed and body weight support at three different settings as done for healthy subjects would take 27 conditions each several minutes long. As patients can usually exercise around 30 to 40 minutes in the Lokomat, the investigations were split up into two patient groups (experiment I and II). Each condition was set to be three minutes long, which was in pre-tests found to be a good tradeoff between experimental time and reaching steady state of heart rate. Ethical approval was obtained by the local ethics committee and all subjects gave informed consent. Although body weight support was not planed to be altered during heart rate control (see Discussion for details), its effects on heart rate were still investigated for theoretical purposes. In a first set of experiments (experiment I), subjects walked at two walking speeds [1.5 and 2.5 km/h], two levels of guidance force [50% and 100%] and two levels of unloading to [30% and 60%]. This resulted in eight conditions plus one baseline condition while standing in the Lokomat. Lower settings for walking speed were reported to feel uncomfortable by patients; 0% GF was not investigated, as only one patient was able to walk at less than 30% guidance force; higher values of BWS were not investigated, as they are not used in a clinical setting. Friction force, as used with healthy subjects for increase of power expenditure, was also not used as it would resist the guidance force provided by the Lokomat.

Table 4.1: Patient data from model identification recordings. (©2009 IEEE)

ID	Injury	Month post incident	Age	Sex	10 m walking [s]	FAC	Func ass.
1	3	44	W	11.7	5	MI leg 83	
2	16	32	W	7.7	5	n.a.	
3	95	48	M	8.7	5	MI leg 33	
4	30	52	M	10.1	5	MI leg 75	
5	14	67	M	7.1	5	MI leg 91	
6	18	61	W	9.1	5	MI leg 75	
7	12	57	M	7.9	5	MI leg 91	
8	99	50	M	6.8	5	MI leg 75	

A second set of experiments was performed (experiment II), to obtain additional data points for different treadmill speeds and unloadings. Investigation of guidance force changes were not included as a result of experiment I, as changes in guidance force did not cause changes in heart rate (see Results section). Treadmill speed was set to [1.5, 2 and 3 km/h] and BWS to [30%, 45% and 60%]. Walking speeds higher than 3 km/h were not investigated for patient safety; lower levels of BWS could not be chosen as the ones of healthy subjects, as no patient was able to walk at 3 km/h with less than 30% BWS. Experiment II had nine conditions plus one baseline condition with condition length set to 3 minutes. All results were computed as percent change in heart rate from baseline standing.

4.3 Results

4.3.1 Model Identification

Healthy subjects did not show an increase of heart rate for changes in BWS or GF, but only for changes in treadmill velocity and friction (Fig 4.3). Therefore, only the influence of changes in treadmill speed and power output were modeled. Heart rate was computed as an increase from baseline

$$HR = HR_{Baseline} + \Delta HR \tag{4.3}$$

with ΔHR = overshoot dynamics + undershoot dynamics + power expenditure + fatigue. The overall heart rate model for healthy subjects was parameterized as

$$\Delta HR = \frac{sk_1 v_{TM}}{s^2 + \tau_{OS}d_{OS} + \tau_{OS}^2} + \frac{sk_2 v_{TM}}{s^2 + \tau_{US}d_{US} + \tau_{US}^2} + \frac{k_3 v_{TM} + k_4 P}{\tau_{fast}s + 1} + \frac{k_5 P}{\tau_{slow}s + 1} \tag{4.4}$$

with P being the power exchanged between human and Lokomat, τ_{OS} and τ_{US} the time constant of overshoot (OS) and undershoot (US) respectively, τ_{fast} and τ_{slow} the time constants of the fast and slow dynamics and $k_i, i \in [1, 4]$ are the scaling factors.

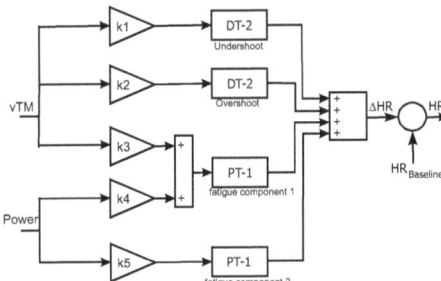

Figure 4.2: Simulink model of the heart rate model components. Inputs are the treadmill speed and the power exchanged between human and Lokomat. Output is the change in heart rate which is added to the baseline heart rate which was recorded during standing. (©2009 IEEE)

The experiments verified the linear dependency between heart rate and power expenditure (Fig. 4.3 A, trendlines) proposed in (Vokac et al., 1975; Borg et al., 1987; Arts and Kuipers, 1994). In the model (eq. 4.4), four of the five scaling factors were found to be subject-depended, the other parameters were distributed within ±10% of their respective mean values and were therefore set to the mean. The constant parameters were $\tau_{slow} = 575.03$, $\tau_{OS} = 0.0575$, $d_{OS} = 1.0094$, $k_{US} = 0.1445$, $d_{US} = 1.0010$, $\tau_{US} = 0.0302$.

4 Monitoring Heart Rate During Control of Physical Effort

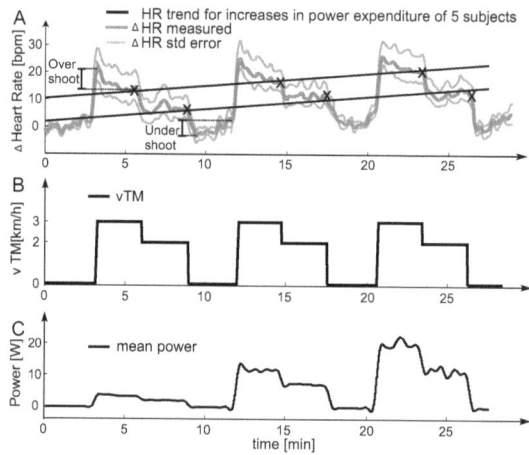

Figure 4.3: A: Changes in heart rate (mean and standard error for five healthy subjects) for different settings of treadmill speed (B) and different levels power expenditure (C). The trend lines show that heart rate in steady state increases linearly with increasing power expenditure. (©2009 IEEE)

4.3.2 Heart Rate Prediction of Healthy Subjects and Patients

The model for treadmill velocity and power expenditure during robot-assisted gait rehabilitation successfully predicted the heart rate of healthy subjects (Fig. 4.4, Tab. 4.4). Whenever healthy subjects did not respond to increases of treadmill speed with increases in heart rate during model identification, a low quality of heart rate prediction was obtained. Particularly the heart rate of subject 4 (Tab. 4.4), an experienced rower, could not be influenced by walking speeds below 3.2 km/h.

Four out of eight patients (Tab. 4.1) did not show changes of heart rate in response to different training conditions. Therefore, patients were classified as responders and none-responders with the following criterion: if the mean heart rate at steady state during the most exhausting condition (3 km/h, 50% guidance force, 30% body weight support) did exceed the mean heart rate of standing by the average heart rate fluctuation (± 4 bpm), then the patient was labeled as responder, otherwise to the nonresponders class. The results of all responders of experiments I and II are summarized in table 4.3. Control in these four subjects is not feasible as there would be no possibility of influencing heart rate with the current setup of the Lokomat.

Table 4.2: Results of changes in guidance force for all patients in experiment I. Percent change in heart rate from baseline, where baseline was set to 100%. (©2009 IEEE)

BWS	v_{TM}	Guidance force [%] 50	100
30	1.5	131 ± 5	130 ± 6
30	2.5	117 ± 3	116 ± 7
60	1.5	115 ± 2	119 ± 3
60	2.5	113 ± 2	111 ± 2

BWS: Body weight support, v_{TM}: treadmill speed

Within the class of responders, a decrease in heart rate was found for decreasing BWS. Contrary to clinical observations informally reported by physiotherapists, no change above or below the normal heart rate fluctuation were found for changes in guidance force (Tab. 4.2).

Table 4.3: Results of changes in for all patients in experiment II. Percent change in heart rate from baseline, where baseline was set to 100%. For some conditions, no measurements were taken. These are indicated by 'n.a.'. (©2009 IEEE)

	Body weight support [%]		
v_{TM}	30	45	60
1	114 ± 1	106 ± 3	103 ± 1
1.5	131 ± 5	n.a.	116 ± 3
2	108 ± 3	107 ± 1	102 ± 2
2.5	117 ± 2	n.a.	114 ± 2
3	112 ± 3	109 ± 1	108 ± 3

v_{TM}: Treadmill speed

Table 4.4: Prediction quality of the heart rate model with optimization of 4 parameters. Subject 1 and 2 were both recorded three times. Subject 4 did not show increases in heart rate above one standard deviation. (©2009 IEEE)

Subject	Recording number	R^2 for optimization over first 12 min	R^2 for 10% power decrease	R^2 for 10% power increase
1	1	0.93	0.91	0.92
	2	0.82	0.80	0.81
	3	0.71	0.70	0.71
2	1	0.92	0.92	0.91
	2	0.91	0.90	0.91
	3	0.85	0.85	0.85
3	1	0.78	0.76	0.77
4	1	0.45	0.44	0.44
5	1	0.85	0.84	0.85

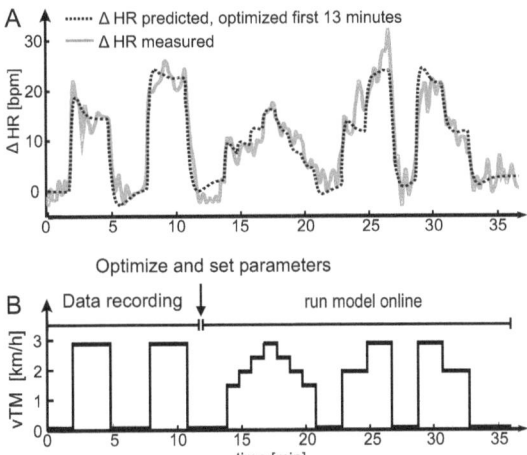

Figure 4.4: Predicted and recorded heart rate of one healthy subject subject (A) and Lokomat treadmill velocity profile (B) for model verification. Heart rate was recorded for the first 12 minutes. From this data, the model parameters were computed and fixed. From minute 12 on, the model predicted heart rate in real-time. (©2009 IEEE)

4.4 Discussion

Decreasing GF and increasing treadmill speed would have been expected to lead to an increase in metabolic cost and, therefore, to increase HR (Borg et al., 1987). Surprisingly, changes in GF did not alter HR in healthy subjects and patients, and HR at treadmill speeds of 1.5 km/h were increased compared to HR at 2.5 km/h in patients (Tab. 4.2). Both results are counter-intuitive.

A possible explanation for the GF results could be that decreased GF did increase metabolic cost, but allowed the subjects a larger step length, as they could deviate from the predefined trajectory. Larger step length could in return have decreased effort and kept the the overall energetic cost of walking constant.

The increase in HR for low gait speeds was informally confirmed by patients, who reported low gait speeds to be more exhausting than faster gait. This might have to do with the effects of leg's vein pumps, which can support the cardiovascular system better at higher gait speeds. Preliminary results of gait speeds up to 4.5 km/h in the Lokomat showed, that HR did increase monotonic between 2 km/h and 4.5 km/h. These higher gait speeds were not available at the time of the experiments.

Not all patients could be classified as responders to our intervention. It was not possible to clearly identify a clinical indicator that would predict pre-training whether or not the patient would react. Beta blocking medication, which is known to decrease HR variability and limit the adaptation of HR to physical stress (Cook et al., 1991), was ruled out as a reason. Post hoc analysis of power expenditure could also not explain the observed phenomena. In stroke survivors, medullar brain stem and hemispheric infarctions were shown to impair autonomic cardiovascular regulation (Korpelainen et al., 1999).

In principle, the model developed in this chapter would be suitable to be used for control purposes. Model Predictive Control could have taken the transfer function from power exchange to heart rate and used it as a disturbance model. Heart rate could then be controlled via adaptation of treadmill speed. In the light of the results of chapter 2, where control of heart rate was possible without model identification, Model Predictive Control would require an unjustifiable amount of preparation effort. Prediction is however possible, as the initial identification period does not interrupt the normal training

4.5 Contribution

The methods presented in this chapter were developed as part of my thesis. Dr. Antonello Caruso and Prof. Dr. Manfred Morari advised me on modeling. The model of power exchange between human and Lokomat was developed in collaboration with my colleague Dr. Heike Vallery.

5 Conclusions and Outlook

5.1 Key Findings

The key result of my studies on control of physiological states is that constantly challenging the patient at his or her maximal capabilities is possible with solutions that are adaptable the patient's respective physical and cognitive capabilities. The use of virtual environments thereby allows providing feedback to patients in an easy to understand way. Physiology based models can be used to prevent overexertion by predicting the temporal evolution of heart rate.

For the first time, control of physical activity was performed

- during robot-assisted gait training,
- in patients with a broad variety of cognitive and motor impairments,
- under model based supervision of heart rate to prevent overexertion.

Besides closed loop physiology control, the role of motivation is known to be important in the success of neurorehabilitation (Robertson and Murre, 1999; Loureiro et al., 2001). The human-in-the-loop structure allows optimizing mental engagement of the subject, thus, increasing motivation. Controlling cognitive engagement in neurorehabilitation as implemented in closed-loop control of psychology, therefore, has the potential to increase motor learning and thereby the training efficiency and therapeutic outcome of neurological rehabilitation (Maclean and Pound, 2000a; Kaelin-Lang et al., 2005).

The key result of my studies on control of psychological states is that real time, objective assessment and control of cognitive load during gait training is possible by using a combination of psychophysiological measurements and task performance as source for state estimation. A major contribution to the day to day clinical applicability of the approach might come from the use of Kalman filters for subject-specific adaptation of classifier parameters. The use of adaptive algorithms for intelligent machine learning as described above could be the basis for future rehabilitation devices that automatically adapt to the specific needs and demands of the patient.

For the first time, closed loop control of cognitive load has been performed

- in neurological patients,
- during execution of a motor task,

5 Conclusions and Outlook

- in the presence of physical effort induce by walking movements.

Detection and control of physiological and psychological states is thereby neither limited to a particular gait orthosis, nor to rehabilitation of the lower limbs. In robot assisted arm rehabilitation, as performed with the ARMin (Nef et al., 2009), the HapticMaster (Houtsma and Van Houten, 2006) or the MIT Manus (Aisen et al., 1997), the lower level of physical effort as compared to walking might even improve the accuracy of the algorithms described above.

5.2 The Impact of Human in the Loop Control on Rehabilitation Robotics

The use of conventional rehabilitation devices can be unsatisfactory, because an efficient interaction between the technical system and the patient is often limited or impossible. Many advanced rehabilitation systems that include novel actuation and digital processing capabilities work in a "master-slave" relationship, thus, tending to force the user only to follow predetermined reference trajectories without taking into account individual properties, spontaneous intentions, or voluntary efforts of that particular person. For instance, many actuated orthoses impose a predetermined motion pattern to the patient's legs, but do not react to the patient's voluntary effort.

In contrast, bio-cooperative rehabilitation technologies offer a new approach by placing the human into the loop, where the human is more than just a sender of the command to a device or the passive receiver of a device action. Placing the human into the loop can be considered from various viewpoints and realized for different applications. It can integrate controlling biomechanical, physiological, as well as psychological aspects of the human, who then represents the plant within the control system. The interaction becomes bi-directional and the technical rehabilitation system takes into account the user's properties, intentions and actions, as well as environmental factors.

In neurorehabilitation, active physical participation was shown to increase motor learning (Lotze et al., 2003). The positive effect of active physical participation on rehabilitation was confirmed by studies that connected cardiovascular training with a positive effect on the recovery after neurological injury (Gordon et al., 2004), exemplary implemented in closed loop heart rate control. Heart rate control in the Lokomat thereby allows cardiovascular training of none ambulatory patients; meanwhile, our applications guarantee that the patient is training in a safe region by keeping relevant physiological values such as heart rate in an appropriate range.

By putting the human into the center of an intelligent control loop, the classical master-slave paradigm can be avoided, which requires the user to adapt to the robotic system. Focusing on integrating bio-cooperative closed loop control based

5.2 The Impact of Human in the Loop Control on Rehabilitation Robotics

on physiological signals reflecting psychological parameters could make the patient the master and the robot the slave. Bio-cooperative control is not limited to the lower extremities. The work described in this thesis was part of the MIMICS project (www.mimics.ethz.ch). Estimation and control of psychological states were also successfully performed in the upper extremities (Novak et al., 2010).

Particularly in the field of gait rehabilitation, the effectiveness of robot intervention is not yet fully realized. The Lokomat gait orthosis provides the advantage of reducing the burden on clinical staff during weight supported gait training and has been found by some researchers to have advantages over conventional therapies (Husemann et al., 2007; Mayr et al., 2007). By addressing important psychological aspects of mental engagement in technology assisted therapies, innovations may be made that help bridge the gap found by some researchers with conventional therapies (Hornby et al., 2008; Hidler et al., 2009). The approach might help to improve some of the features of technology assisted rehabilitation - taking a step towards meeting more of the needs of the patient and therapist.

In summary, closed loop control of mental states has the potential to improve robot assisted rehabilitation by enabling clinicians to provide patient-centered rehabilitation therapy. In the future, human-in-the-loop strategies will break with the classical master-slave paradigm that requires the user to adapt to the robotic system. Focusing on integrating mental states in the control loop will make the patient the master and the robot the slave. By using auto-adaptive algorithms such as intelligent machine learning as described above, the robot will learn how to automatically adapt to the specific needs and demands of the patient.

Bibliography

Achten, J. and Jeukendrup, A. E. (2003). Heart rate monitoring: applications and limitations. *Sports Med*, 33(7):517–38. Achten, Juul Jeukendrup, Asker E Review New Zealand Sports medicine (Auckland, N.Z.) Sports Med. 2003;33(7):517-38.

Adams, R., Moreyra, M., and Hannaford, B. (1998). Stability and performance of haptic displays: Theory and experiments. *Proceedings ASME Int. Mech. Engineering Congress and Exhibition*, page 8.

Adams, R. J. and Hannaford, B. (1999). Stable haptic interaction with virtual environments. *Robotics and Automation, IEEE Transactions on*, 15(3):465–474.

Aisen, M. L., Krebs, H. I., Hogan, N., McDowell, F., and Volpe, B. T. (1997). The effect of robot-assisted therapy and rehabilitative training on motor recovery following stroke. *Arch Neurol*, 54(4):443–6. Aisen, M L Krebs, H I Hogan, N McDowell, F Volpe, B T Research Support, Non-U.S. Gov't United states Archives of neurology Arch Neurol. 1997 Apr;54(4):443-6.

Anderson, K. D. (2004). Targeting recovery: priorities of the spinal cord-injured population. *J Neurotrauma*, 21(10):1371–83.

Andreassi, J. (2007). *Psychophysiology: Human behavior and physiological response*. Lawrence Erlbaum Associates, Mahwah, 5th edition.

Arts, F. J. and Kuipers, H. (1994). The relation between power output, oxygen uptake and heart rate in male athletes. *Int J Sports Med*, 15(5):228–31. Arts, F J Kuipers, H Germany International journal of sports medicine Int J Sports Med. 1994 Jul;15(5):228-31.

Ax, A. F. (1953). The physiological differentiation between fear and anger in humans. *Psychosomatic Medicine*, 15(5):433–442.

Banz, R., Bolliger, M., Colombo, G., Dietz, V., and Lunenburger, L. (2008). Computerized visual feedback: an adjunct to robotic-assisted gait training. *Phys Ther*, 88(10):1135–45.

Basdogan, C., Ho, C. H., and Srinivasan, M. A. (2001). Virtual environments for medical training: graphical and haptic simulation of laparoscopic common bile duct exploration. *Mechatronics, IEEE/ASME Transactions on*, 6(3):269–285.

Bibliography

Baum, K., Essfeld, D., Leyk, D., and Stegemann, J. (1992). Blood pressure and heart rate during rest-exercise and exercise-rest transitions. *Eur J Appl Physiol Occup Physiol*, 64(2):134–8. Baum, K Essfeld, D Leyk, D Stegemann, J Germany European journal of applied physiology and occupational physiology Eur J Appl Physiol Occup Physiol. 1992;64(2):134-8.

Bavelier, D., Levi, D. M., Li, R. W., Dan, Y., and Hensch, T. K. (2010). Removing brakes on adult brain plasticity: From molecular to behavioral interventions. *Journal of Neuroscience*, 30(45):14964–14971.

Bernardi, L., Valle, F., Coco, M., Calciati, A., and Sleight, P. (1996). Physical activity influences heart rate variability and very-low-frequency components in holter electrocardiograms. *Cardiovasc Res*, 32(2):234–7.

Bishop, C. (2006). *Pattern Recognition and Machine Learning*. Information Science and Statistics. Springer.

Boiten, F. A., Frijda, N. H., and Wientjes, C. J. E. (1994). Emotions and respiratory patterns: review and critical analysis. *International Journal of Psychophysiology*, 17(2):103–128.

Borg, G., Hassmen, P., and Lagerstrom, M. (1987). Perceived exertion related to heart rate and blood lactate during arm and leg exercise. *Eur J Appl Physiol Occup Physiol*, 56(6):679–85. Borg, G Hassmen, P Lagerstrom, M Research Support, Non-U.S. Gov't Germany, west European journal of applied physiology and occupational physiology Eur J Appl Physiol Occup Physiol. 1987;56(6):679-85.

Boucsein, W. (2005). Electrodermal measurement. In Stanton, N., Hedge, A., Brookhuis, K., Salas, E., and Hendrick, H., editors, *Handbook of human factors and ergonomics methods*, pages 18–1–18–8. CRC Press, London.

Brainin, M., Bornstein, N., Boysen, G., and Demarin, V. (2000). Acute neurological stroke care in europe: results of the european stroke care inventory. *Eur J Neurol*, 7(1):5–10.

Brown, S.-L. and Schwartz, G. E. (1980). Relationships between facial electromyography and subjective experience during affective imagery. *Biological Psychology*, 11(1):49–62.

Brütsch, K., Schuler, T., Koenig, A., Zimmerli, L., Merillat-Koeneke, S., Lunenburger, L., Riener, R., Jancke, L., and Meyer-Heim, A. (2010). Influence of virtual reality soccer game on walking performance in robotic assisted gait training for children. *J Neuroeng Rehabil*, 7(1):15.

Bibliography

Buttolo, P., Oboe, R., and Hannaford, B. (1997). Architectures for shared haptic virtual environments. *Computers and Graphics*, 21(4):421–429.

Cacioppo, J. T., Tassinary, L., and Berntson, G. (2000). *Handbook of Psychophysiology*. Cambridte Press 2000, 2nd edition.

Carroll, D., Turner, J. R., and Prasad, R. (1986b). The effects of level of difficulty of mental arithmetic challenge on heart rate ande oxygen-consumption. *International Journal of Psychophysiology*, 4(3):167–173.

Cheng, T., A, S., Su, S., Celler, B., and Wang, L. (2008). A robust control design for heart rate tracking during exercise.

Cherng, R. J., Liu, C. F., Lau, T. W., and Hong, R. B. (2007). Effect of treadmill training with body weight support on gait and gross motor function in children with spastic cerebral palsy. *Am J Phys Med Rehabil*, 86(7):548–55.

Christov (2004). Real time electrocardiogram qrs detection using combined adaptive threshold. *BioMedical Engineering OnLine*, 3(28):8.

Codispoti, M., Surcinelli, P., and Baldaro, B. (2008). Watching emotional movies: Affective reactions and gender differences. *International Journal of Psychophysiology*, 69(2):90–95.

Colgate, E. and Hogan, N. (1989). *The Interaction of Robots with Passive Environments: Application to Force Feedback Control*. In Advanced Robotics. Springer-Verlag, Berlin.

Colombo, G., Joerg, M., Schreier, R., and Dietz, V. (2000). Treadmill training of paraplegic patients using a robotic orthosis. *J Rehabil Res Dev*, 37(6):693–700.

Colombo, R., Pisano, F., Mazzone, A., Delconte, C., Micear, S., Carrozza, M., Dario, P., and Minuco, G. (2007). Design strageties to improve patient motivation during robot-aided rehabilitation. *Journal of NeuroEngineering and Rehabilitation*, 4(3):12.

Cook, J. R., Bigger, J. T., J., Kleiger, R. E., Fleiss, J. L., Steinman, R. C., and Rolnitzky, L. M. (1991). Effect of atenolol and diltiazem on heart period variability in normal persons. *J Am Coll Cardiol*, 17(2):480–4.

Dawson, M. E., Schell, A. M., and Filion, D. L. (2007). *Handbook of Psychophysiology*. 3rd edition.

de la Cruz Torres, B., Lopez, C. L., and Orellana, J. N. (2008). Analysis of heart rate variability at rest and during aerobic exercise: a study in healthy people and cardiac patients. *British Journal of Sports Medicine*, 42(9):715–720.

Bibliography

Delaney, J. P. A. and Brodie, D. A. (2000). Effects of short-term psychological stress on the time and frequency domains of heart-rate variability. *Perceptual and Motor Skills*, 91(2):515–524.

Dietz, V. and Duysens, J. (2000). Significance of load receptor input during locomotion: a review. *Gait and Posture*, 11:9.

Dietz, V., Wirz, M., Curt, A., and Colombo, G. (1998). Locomotor pattern in paraplegic patients: training effects and recovery of spinal cord function. *Spinal Cord*, 36(6):380–90.

Duffy, V. G. (2008). *Handbook of Digital Human Modeling: Research for Applied Ergonomics and Human Factors Engineering*. CRC Press, Inc.

Duschau-Wicke, A., von Zitzewitz, J., Caprez, A., Lunenburger, L., and Riener, R. (2010). Path control: A method for patient-cooperative robot-aided gait rehabilitation. *IEEE Trans Neural Syst Rehabil Eng*, 18:38–48.

Fagraeus, L. and Linnarsson, D. (1976). Autonomic origin of heart rate fluctuations at the onset of muscular exercise. *J Appl Physiol*, 40(5):679–82. Fagraeus, L Linnarsson, D United states Journal of applied physiology J Appl Physiol. 1976 May;40(5):679-82.

Feroldi, P., Belleri, M., Ferretti, G., and Veicsteinas, A. (1992). Heart rate overshoot at the beginning of muscle exercise. *Eur J Appl Physiol Occup Physiol*, 65(1):8–12. Feroldi, P Belleri, M Ferretti, G Veicsteinas, A Research Support, Non-U.S. Gov't Germany European journal of applied physiology and occupational physiology Eur J Appl Physiol Occup Physiol. 1992;65(1):8-12.

Fitts, P. M. (1954). The information capacity of the human motor system in controlling the amplitude of movement. *Journal of experimental psychology*, 47(6):381–91.

Folstein, M. F., Folstein, S. E., and McHugh, P. R. (1975). 'mini-mental state'. a practical method for grading the cognitive state of patients for the clinician. *J Psychiatr Res*, 12(3):189–98.

Fong, T., Thorpe, C., and Baur, C. (2003). Collaboration, dialogue, human-robot interaction. In Jarvis, R. and Zelinsky, A., editors, *Robotics Research*, volume 6 of *Springer Tracts in Advanced Robotics*, pages 255–266. Springer Berlin / Heidelberg.

Fung, J., Richards, C. L., Malouin, F., McFadyen, B. J., and Lamontagne, A. (2006). A treadmill and motion coupled virtual reality system for gait training post-stroke. *Cyberpsychol Behav*, 9(2):157–62.

Girone, M., Burdea, G., Bouzit, M., Popescu, V., and Deutsch, J. E. (2000). Orthopedic rehabilitation using the "rutgers ankle" interface. *Stud Health Technol Inform*, 70:89–95.

Gomez, P., Zimmermann, P., Guttormsen-Schr, S., and Danuser, B. (2005). Respiratory responses associated with affective processing of film stimuli. *Biological Psychology*, 68(3):223–235.

Gordon, N. F., Gulanick, M., Costa, F., Fletcher, G., Franklin, B. A., Roth, E. J., and Shephard, T. (2004). Physical activity and exercise recommendations for stroke survivors: an american heart association scientific statement from the council on clinical cardiology, subcommittee on exercise, cardiac rehabilitation, and prevention; the council on cardiovascular nursing; the council on nutrition, physical activity, and metabolism; and the stroke council. *Stroke*, 35(5):1230–40.

Guadagnoli, M. A. and Lee, T. D. (2004). Challenge point: a framework for conceptualizing the effects of various practice conditions in motor learning. *J Mot Behav*, 36(2):212–24.

Guzella, L. (2007). *Anaysis and Synthesis of Single-Input Single-Output Control Systems*. vdf Hochschulverlag AG, Zurich, Switzerland.

Haarmann, A., Boucsein, W., and Schaefer, F. (2009). Combining electrodermal responses and cardiovascular measures for probing adaptive automation during simulated flight. *Applied Ergonomics*, 40(6):1026–1040.

Hajek, M., Potucek, J., and Brodan, V. (1980). Mathematical-model of heart rate regulation during exercise. *Automatica*, 16(2):5.

Hesse, S., Sarkodie-Gyan, T., and Uhlenbrock, D. (1999). Development of an advanced mechanised gait trainer, controlling movement of the centre of mass, for restoring gait in non-ambulant subjects. *Biomed Tech (Berl)*, 44(7-8):194–201.

Hidler, J., Nichols, D., Pelliccio, M., Brady, K., Campbell, D. D., Kahn, J. H., and Hornby, T. G. (2009). Multicenter randomized clinical trial evaluating the effectiveness of the lokomat in subacute stroke. *Neurorehabil Neural Repair*, 23(1):5–13.

Hidler, J. M. and Wall, A. E. (2005). Alterations in muscle activation patterns during robotic-assisted walking. *Clin Biomech (Bristol, Avon)*, 20(2):184–93.

Hogan, N. (1985). Impedance control: An approach to manipulation. *J Dyn Syst-T ASME*, 107:1–23.

Holden, M. K. (2005). Virtual environments for motor rehabilitation: review. *Cyberpsychol Behav*, 8(3):187–211; discussion 212–9.

Holmgren, A. (1956). Circulatory changes during muscular work in man; with special reference to arterial and central venous pressures in the systemic circulation. *Scand J Clin Lab Invest*, 8(Suppl 24):1–97. Holmgren, a Not Available Scandinavian journal of clinical and laboratory investigation Scand J Clin Lab Invest. 1956;8(Suppl 24):1-97.

Hornby, T. G., Campbell, D. D., Kahn, J. H., Demott, T., Moore, J. L., and Roth, H. R. (2008). Enhanced gait-related improvements after therapist- versus robotic-assisted locomotor training in subjects with chronic stroke: a randomized controlled study. *Stroke*, 39(6):1786–92.

Houtsma, J. A. and Van Houten, F. J. (2006). Virtual reality and a haptic master-slave set-up in post-stroke upper-limb rehabilitation. *Proc Inst Mech Eng H*, 220(6):715–8. Houtsma, J A Van Houten, F J A M England Proceedings of the Institution of Mechanical Engineers. Part H, Journal of engineering in medicine Proc Inst Mech Eng H. 2006 Aug;220(6):715-8.

Hugdahl, K. (1995). *Psychophysiology: The Mind-body Perspective*. Harvard University Press.

Husemann, B., Muller, F., Krewer, C., Heller, S., and Koenig, E. (2007). Effects of locomotion training with assistance of a robot-driven gait orthosis in hemiparetic patients after stroke: a randomized controlled pilot study. *Stroke*, 38(2):349–54.

Israel, J. F., Campbell, D. D., Kahn, J. H., and Hornby, T. G. (2006). Metabolic costs and muscle activity patterns during robotic- and therapist-assisted treadmill walking in individuals with incomplete spinal cord injury. *Phys Ther*, 86(11):1466–78.

Johnson, M. (2006). Recent trends in robot-assisted therapy environments to improve real-life functional performance after stroke. *J Neuroeng Rehabil*, 3:29.

Kaelin-Lang, A., Sawaki, L., and Cohen, L. G. (2005). Role of voluntary drive in encoding an elementary motor memory. *J Neurophysiol.*, 93(2):1099–1103.

Koenig, A., Novak, D., Pulfer, M., Omlin, X., Perreault, E., Zimmerli, L., Mihelj, M., and Riener, R. (2011a). Real-time closed-loop control of cognitive load in neurological patients during robot-assisted gait training. *Trans Neural Sys Rehab Eng*, 19:453–464.

Koenig, A., Omlin, X., Bergmann, J., Zimmerli, L., Bolliger, M., Mueller, F., and Riener, R. (2011b). Controlling patient participation during robot-assisted gait training. *J Neuroeng Rehabil*, 8.

Koenig, A., Omlin, X., Domen, N., and Riener, R. (2011c). A review on biocooperative control in gait rehabilitation. In *Int. Conf. Rehab. Robotics (ICORR) 2011*, page 6, Zurich, Switzerland.

Koenig, A., Omlin, X., Zimmerli, L., and Riener, R. (2011d). Virtual environments in neurological gait rehabilitation for automated control of physical activity. In *Technically Asssisted Rehabilitati (TAR) 2011*, page 2, Berlin, Germany.

Koenig, A., Omlin, X., Zimmerli, L., Sapa, M., Krewer, C., Bolliger, M., Mueller, F., and Riener, R. (2011e). Psychological state estimation from physiological recordings during robot assisted gait rehabilitation. *JRRD*, 48(4).

Koenig, A., Pulfer, M., Omlin, X., Perreault, E., Zimmerli, L., and Riener, R. (2010). Automatic estimation of cognitive load during robot-assisted gait training. In *Automed Workshop*, page 2, Zurich, Switzerland.

Koenig, A. C., Somaini, L., Pulfer, M., Holenstein, T., Omlin, X., Wieser, M., and Riener, R. (2009). Model-based heart rate prediction during lokomat walking. *Conf Proc IEEE Eng Med Biol Soc*, 2009:1758–61.

Korpelainen, J., Tolonen, U., Sotaniemi, K., and Myllyla, V. (1993). Suppressed sympathetic skin response in brain infarction. *Stroke*, 24(9):1389–1392.

Korpelainen, J. T., Sotaniemi, K. A.Makikallio, A., Huikuri, H. V., and Myllyla, V. V. (1999). Dynamic behavior of heart rate in ischemic stroke. *Stroke*, 30(5):1008–13.

Korpelainen, J. T., Sotaniemi, K. A., Huikuri, H. V., and Myllyla, V. V. (1996). Abnormal heart rate variability as a manifestation of autonomic dysfunction in hemispheric brain infarction. *Stroke*, 27(11):2059–2063.

Kwakkel, G., Kollen, B. J., and Wagenaar, R. C. (2002). Long term effects of intensity of upper and lower limb training after stroke: a randomised trial. *J Neurol Neurosurg Psychiatry*, 72(4):473–9.

Kwakkel, G., Wagenaar, R. C., Koelman, T. W., Lankhorst, G. J., and Koetsier, J. C. (1997). Effects of intensity of rehabilitation after stroke. a research synthesis. *Stroke*, 28(8):1550–6.

Langhorne, P., Wagenaar, R., and Partridge, C. (1996). Physiotherapy after stroke: more is better? *Physiother Res Int*, 1(2):75–88.

Larsen, J. T., Norris, C. J., and Cacioppo, J. T. (2003). Effects of positive and negative affect on electromyographic activity over zygomaticus major and corrugator supercilii. *Psychophysiology*, 40:776–785.

Levenson, R., Ekman, P., and Friesen, W. (1990). Voluntary facial action generates emotion-specific autonomic nervous system activity. *Psychophysiology*, 27(4):363–384.

Liebermann, D. G., Buchman, A. S., and Franks, I. M. (2006). Enhancement of motor rehabilitation through the use of information technologies. *Clin Biomech (Bristol, Avon)*, 21(1):8–20.

Liu, C. C., Agrawal, P., Sarkar, N., and Chen, S. O. (2009). Dynamic difficulty adjustment in computer games through real-time anxiety-based affective feedback. *International Journal of Human-Computer Interaction*, 25(6):506–529.

Lotze, M., Braun, C., Birbaumer, N., Anders, S., and Cohen, L. G. (2003). Motor learning elicited by voluntary drive. *Brain*, 126(Pt 4):866–72.

Loureiro, R., Amirabdollahian, F., S. Cootes, S., Stokes, E., and Harwin, W. (2001). Using haptics technology to deliver motivational therapies in stroke patients: Concepts and initial pilot studies. *Proceedings of EuroHaptics 2001*, page 6.

Lunenburger, L., Colombo, G., and Riener, R. (2007). Biofeedback for robotic gait rehabilitation. *J Neuroengineering Rehabil*, 4:1.

Lunenburger, L., Colombo, G., Riener, R., and Dietz, V. (2004). Biofeedback in gait training with the robotic orthosis lokomat. *Conf Proc IEEE Eng Med Biol Soc*, 7:4888–91.

MacKenzie, I. S. (1995). Movement time prediction in human-computer interfaces. In Baecker, R. M., Buxton, W. A. S., Grudin, J., and Greenberg, S., editors, *Readings in human-computer interaction*, pages 483–493. Kaufmann, Los Altos, CA, USA.

Maclean, N. and Pound, P. (2000a). A critical review of the concept of patient motivation in the literature on physical rehabilitation. *Soc Sci Med*, 50(4):495–506.

Maclean, N. and Pound, P. (2000b). A critical review of the concept of patient motivation in the literature on physical rehabilitation. *Soc Sci Med*, 50(4):495–506.

Malik, M. (1996). Heart rate variability - standards of measurement, physiological interpretation and clinical use. *European Heart Journal*, 17:28.

Mancuso, D. L. and Knight, K. L. (1992). Effects of prior physical activity on skin surface temperature response of the ankle during and after a 30-minute ice pack application. *Journal of Athletic Training*, 27(3):242, 244, 246, 248–249.

Mayr, A., Kofler, M., Quirbach, E., Matzak, H., Frohlich, K., and Saltuari, L. (2007). Prospective, blinded, randomized crossover study of gait rehabilitation in stroke patients using the lokomat gait orthosis. *Neurorehabil Neural Repair*, 21(4):307–14.

McAuley, E., Duncan, T., and Tammen, V. V. (1989). Psychometric properties of the intrinsic motivation inventory in a competitive sports setting - a confirmatory factor analysis. *Research Quarterly for Exercise and Sport*, 60(1):48–58.

McCraty, R., Atkinson, M., Tiller, W. A., Rein, G., and Watkins, A. D. (1995). The effects of emotions on short-term power spectrum analysis of heart-rate-variability. *American Journal of Cardiology*, 76(14):1089–1093.

Mehrholz, J., Kugler, J., and Pohl, M. (2008). Locomotor training for walking after spinal cord injury. *Cochrane Database Syst Rev*, (2):CD006676.

Mehrholz, J., Werner, C., Kugler, J., and Pohl, M. (2007). Electromechanical-assisted training for walking after stroke. *Cochrane Database Syst Rev*, (4):CD006185.

Meyer, S., Strittmatter, M., Fischer, C., Georg, T., and Schmitz, B. (2004). Lateralization in autonomic dysfunction in ischemic stroke involving the insular cortex. *NeuroReport*, 15(2):357–361.

Mirelman, A., Bonato, P., and Deutsch, J. E. (2009). Effects of training with a robot-virtual reality system compared with a robot alone on the gait of individuals after stroke. *Stroke*, 40(1):169–74.

Morris, J. D. (1995). Observations: Sam: The self-assessment manikin - an efficient cross-cultural measurement of emotional response. *Journal of Advertising Research*, 35(6):63–68.

Mulder, G., Mulder, L., Veldman, J., and van Roon, A. (2000). A psychophysiological approach to working conditions. In Backs, R. and Boucsein, W., editors, *Engineering psychophysiology: Issues and applications*, pages 139–159. Lawrence Erlbaum Associates, Mahwah.

Naver, H., Blomstrand, C., Ekholm, S., Jensen, C., Karlsson, T., and Wallin, B. G. (1995). Autonomic and thermal sensory symptoms and dysfunction after stroke. *Stroke*, 26(8):1379–1385.

Nef, T., Guidali, M., and Riener, R. (2009). Armin iii - arm therapy exoskeleton with an ergonomic shoulder actuation. *Appl Bionics Biomech*, 6(2):16.

Nef, T., Mihelj, M., and Riener, R. (2007). Armin: a robot for patient-cooperative arm therapy. *Med Biol Eng Comput*, 45(9):887–900.

Novak, D., Ziherl, J., Olensek, A., Milavec, M., Podobnik, J., Mihelj, M., and Munih, M. (2010). Psychophysiological responses to robotic rehabilitation tasks in stroke. *IEEE Trans Neural Syst Rehabil Eng*, 18(4):351–61.

Nudo, R. J. (2006). Plasticity. *NeuroRx*, 3(4):420–7.

O'Gorman, G. (1975). Anti-motivation. *Physiotherapy*, 61(6):176–9. Journal Article England.

Ohsuga, M., Shimono, F., and Genno, H. (2001). Assessment of phasic work stress using autonomic indices. *International Journal of Psychophysiology*, 40(3):211–220.

Ohsuga, M., Tatsuno, Y., Shimono, F., Hirasawa, K., Oyama, H., and Okamura, H. (1998). Bedside wellness–development of a virtual forest rehabilitation system. *Stud Health Technol Inform*, 50:168–74.

Palomba, D., Sarlo, M., Angrilli, A., Mini, A., and Stegagno, L. (2000). Cardiac responses associated with affective processing of unpleasant film stimuli. *International Journal of Psychophysiology*, 36(1):45–57.

Pennycott, A., Hunt, K., Coupaud, S., Allan, D., and Kakebeeke, T. (2010). Feedback control of oxygen uptake during robot-assisted gait. *IEEE Trans Control Systems Tech*, 18(1):7.

Phillips, E. M., Schneider, J. C., and Mercer, G. R. (2004). Motivating elders to initiate and maintain exercise. *Arch Phys Med Rehabil*, 85(7 Suppl 3):S52–7; quiz S58–9.

Pierpont, G. L., Stolpman, D. R., and Gornick, C. C. (2000). Heart rate recovery post-exercise as an index of parasympathetic activity. *J Auton Nerv Syst*, 80(3):169–74. Pierpont, G L Stolpman, D R Gornick, C C Clinical Trial Research Support, U.S. Gov't, Non-P.H.S. Netherlands Journal of the autonomic nervous system J Auton Nerv Syst. 2000 May 12;80(3):169-74.

Rani, P., Sims, J., Brackin, R., and Sarkar, N. (2002). Online stress detection using psychophysiological signals for implicit human-robot cooperation. *Robotica*, 20:13.

Redding, G. M., Rader, S. D., and Lucas, D. R. (1992). Cognitive load and prism adaptation. *Journal of Motor Behavior*, 24(3):238 – 246.

Riener, R., Duschau-Wicke, A., Koenig, A., Bolliger, M., Wieser, M., and Vallery, H. (2009a). Automation in rehabilitation: How to include the human into the loop. In Dssel, O. and Schlegel, W. C., editors, *World Congress on Medical*

Physics and Biomedical Engineering, September 7 - 12, 2009, Munich, Germany, volume 25/13 of *IFMBE Proceedings*, pages 180–183. Springer Berlin Heidelberg.

Riener, R., Frey, M., Bernhardt, M., Nef, T., and Colombo, G. (2005a). Human-centered rehabilitation robotics. In *ICORR*, Chicago.

Riener, R., Koenig, A., Bolliger, M., Wieser, M., Duschau-Wicke, A., and Vallery, H. (2009b). Bio-cooperative robotics: controlling mechanical, physiological and psychological patient states. In *ICORR*, Kyoto.

Riener, R., Lunenburger, L., Jezernik, S., Anderschitz, M., Colombo, G., and Dietz, V. (2005b). Patient-cooperative strategies for robot-aided treadmill training: first experimental results. *IEEE Trans Neural Syst Rehabil Eng*, 13(3):380–94.

Riener, R., Lunenburger, L., Maier, I., Colombo, G., and Dietz, V. (2010). Locomotor training in subjects with sensori-motor deficits: An overview over the robotic gait orthosis lokomat. *J Healthcare Eng*, 1(2):20.

Riener, R. and Munih, M. (2010). Guest editorial: Special section on rehabilitation via bio-cooperative control. *IEEE Trans Neural Syst Rehabil Eng*, 18(4):2.

Robertson, I. H. and Murre, J. M. (1999). Rehabilitation of brain damage: Brain plasticity and prinicples of guided recovery. *Psychological Bulletin*, 125(32):544.

Rousselle, J. G., Blascovich, J., and Kelsey, R. M. (1995). Cardiorespiratory response under combined psychological and exercise stress. *International Journal of Psychophysiology*, 20(1):49–58.

Russell, J. (1980). A circumplex model of affect. *J of Personality and Social Psychology*, 39(6):18.

Sage, G. (1977). *Introduction to motor behavior: A neuropsychological approach*. Addison-Wesley, 2 edition.

Schmidt, H., Werner, C., Bernhardt, R., Hesse, S., and Kruger, J. (2007). Gait rehabilitation machines based on programmable footplates. *Journal of Neuro-Engineering and Rehabilitation*, 4(1):2.

Schwartz, I., Sajin, A., Fisher, I., Neeb, M., Shochina, M., Katz-Leurer, M., and Meiner, Z. (2009). The effectiveness of locomotor therapy using robotic-assisted gait training in subacute stroke patients: A randomized controlled trial. *PM and R*, 1(6):516–523.

Sekhon, L. H. and Fehlings, M. G. (2001). Epidemiology, demographics, and pathophysiology of acute spinal cord injury. *Spine*, 26(24S):S2–S12.

Shumway-Cook, A. and Woollacott, M. (1995). *Motor Control: Theory and Practical Applications*. Lippincott Williams and Wilkins.

Stauffer, Y., Allemand, Y., Bouri, M., Fournier, J., Clavel, R., Metrailler, P., Brodard, R., and Reynard, F. (2009). The walktrainer–a new generation of walking reeducation device combining orthoses and muscle stimulation. *IEEE Trans Neural Syst Rehabil Eng*, 17(1):38–45.

Su, S., Huang, S., Wang, L., Celler, B., Savkin, A., Guo, Y., and Cheng, T. (2007a). Nonparametric hammerstein model based model predictive control for heart rate regulation. In *IEEE EMBS*, Lyon, France.

Su, S. W., Wang, L., Celler, B. G., Savkin, A. V., and Guo, Y. (2007b). Identification and control for heart rate regulation during treadmill exercise. *IEEE Trans Biomed Eng*, 54(7):1238–46.

Suess, W. M., Alexander, A. B., Smith, D. D., Sweeney, H. W., and Marion, R. J. (1980). The effects of psychological stress on respiration - a preliminary-study of anxiety and hyperventilation. *Psychophysiology*, 17(6):535–540.

Sveistrup, H. (2004). Motor rehabilitation using virtual reality. *J Neuroeng Rehabil*, 1(1):10.

Taylor, J. A. and Thoroughman, K. A. (2008). Motor adaptation scaled by the difficulty of a secondary cognitive task. *PLoS One*, 3(6):e2485.

Thomas, E. E., De Vito, G., and Macaluso, A. (2007). Physiological costs and temporo-spatial parameters of walking on a treadmill vary with body weight unloading and speed in both healthy young and older women. *Eur J Appl Physiol*, 100(3):293–9.

Thorvaldsen, P., Asplund, K., Kuulasmaa, K., Rajakangas, A. M., and Schroll, M. (1995). Stroke incidence, case fatality, and mortality in the who monica project. world health organization monitoring trends and determinants in cardiovascular disease. *Stroke*, 26(3):361–7.

Tokgozoglu, S. L., Batur, M. K., Topcuoglu, M. A., Saribas, O., Kes, S., and Oto, A. (1999). Effects of stroke localization on cardiac autonomic balance and sudden death. *Stroke*, 30(7):1307–1311.

van der Meijden, O. and Schijven, M. (2009). The value of haptic feedback in conventional and robot-assisted minimal invasive surgery and virtual reality training: a current review. *Surgical Endoscopy*, 23(6):1180–1190.

Veneman, J. F., Kruidhof, R., Hekman, E. E., Ekkelenkamp, R., Van Asseldonk, E. H., and van der Kooij, H. (2007). Design and evaluation of the lopes exoskeleton robot for interactive gait rehabilitation. *IEEE Trans Neural Syst Rehabil Eng*, 15(3):379–86.

Vidaurre, C., Schlogl, A., Cabeza, R., Scherer, R., and Pfurtscheller, G. (2007). Study of on-line adaptive discriminant analysis for eeg-based brain computer interfaces. *IEEE Trans Biomed Eng*, 54(3):550–6.

Vokac, Z., Bell, H., Bautz-Holter, E., and Rodahl, K. (1975). Oxygen uptake/heart rate relationship in leg and arm exercise, sitting and standing. *J Appl Physiol*, 39(1):54–9. Vokac, Z Bell, H Bautz-Holter, E Rodahl, K United states Journal of applied physiology J Appl Physiol. 1975 Jul;39(1):54-9.

Wellner, M., Schaufelberger, A., Zitzewitz, J. v., and Riener, R. (2008). Evaluation of visual and auditory feedback in virtual obstacle walking. *Presence: Teleoperators and Virtual Environments*, 17(5):512–24.

Westlake, K. P. and Patten, C. (2009). Pilot study of lokomat versus manual-assisted treadmill training for locomotor recovery post-stroke. *J Neuroeng Rehabil*, 6:18.

Wilson, G. F. and Russell, C. A. (2007). Performance enhancement in an uninhabited air vehicle task using psychophysiologically determined adaptive aiding. *Human Factors*, 49(6):1005–1018.

Wirz, M., Zemon, D. H., Rupp, R., Scheel, A., Colombo, G., Dietz, V., and Hornby, T. G. (2005). Effectiveness of automated locomotor training in patients with chronic incomplete spinal cord injury: a multicenter trial. *Arch Phys Med Rehabil*, 86(4):672–80.

Wyndaele, M. and Wyndaele, J. J. (2006). Incidence, prevalence and epidemiology of spinal cord injury: what learns a worldwide literature survey? *Spinal Cord*, 44(9):523–9.

Zimmerli, L., Duschau-Wicke, A., Mayr, A., Riener, R., and Lunenburger, L. (2009). In *ICORR*, pages 150–153, Kyoto, Japan.

Die VDM Verlagsservicegesellschaft sucht für wissenschaftliche Verlage abgeschlossene und herausragende

Dissertationen, Habilitationen, Diplomarbeiten, Master Theses, Magisterarbeiten usw.

für die kostenlose Publikation als Fachbuch.

Sie verfügen über eine Arbeit, die hohen inhaltlichen und formalen Ansprüchen genügt, und haben Interesse an einer honorarvergüteten Publikation?

Dann senden Sie bitte erste Informationen über sich und Ihre Arbeit per Email an *info@vdm-vsg.de*.

Sie erhalten kurzfristig unser Feedback!

VDM Verlagsservicegesellschaft mbH
Dudweiler Landstr. 99
D - 66123 Saarbrücken

Telefon +49 681 3720 174
Fax +49 681 3720 1749

www.vdm-vsg.de

Die VDM Verlagsservicegesellschaft mbH vertritt

Printed by Books on Demand GmbH, Norderstedt / Germany